AF358700

Novel Techniques to Measure the Sensory, Emotional, and Physiological (Biometric) Responses of Consumers toward Foods and Packaging

Novel Techniques to Measure the Sensory, Emotional, and Physiological (Biometric) Responses of Consumers toward Foods and Packaging

Editor

Damir Torrico

MDPI • Basel • Beijing • Wuhan • Barcelona • Belgrade • Manchester • Tokyo • Cluj • Tianjin

Editor
Damir Torrico
Department of Wine, Food and
Molecular Biosciences
Faculty of Agriculture and Life
Sciences
Lincoln University
Lincoln
New Zealand

Editorial Office
MDPI
St. Alban-Anlage 66
4052 Basel, Switzerland

This is a reprint of articles from the Special Issue published online in the open access journal *Foods* (ISSN 2304-8158) (available at: www.mdpi.com/journal/foods/special_issues/Sensory_Emotional_ Physiological_Responses_Consumers_Foods_Packaging).

For citation purposes, cite each article independently as indicated on the article page online and as indicated below:

LastName, A.A.; LastName, B.B.; LastName, C.C. Article Title. *Journal Name* **Year**, *Volume Number*, Page Range.

ISBN 978-3-0365-2537-2 (Hbk)
ISBN 978-3-0365-2536-5 (PDF)

Contents

About the Editor

Damir Torrico

Damir Dennis Torrico holds a PhD in Food Science from Louisiana State University, USA. He is currently a Senior Lecturer in Sensory Science at Lincoln University, New Zealand. His research interests are focused on sensory science in the area of foods and beverages. He has published several research papers in taste perception, food product optimization, consumer acceptability, and physiological/psychological responses of consumers toward food products. He is also the Director of the Centre of Excellence –Food for Future Consumers at Lincoln University, which focuses on interdisciplinary research in the area of food science. Currently, Dr. Torrico has published 71 refereed articles (with an h-index of 21, and an i10-index of 40) with a total of 1247 citations (until 2021) in the area of food and sensory sciences.

Preface to "Novel Techniques to Measure the Sensory, Emotional, and Physiological (Biometric) Responses of Consumers toward Foods and Packaging"

Sensory analysis is used to evaluate the quality and acceptability of foods, beverages, and packaging. However, sensory tests rely on the analysis of self-reported responses to measure the various sensory parameters, and self-reported inputs are generally affected by cognitive biases, which influence the decision-making around the evaluated products. Novel measurement techniques including biometrics based on non-invasive instruments and video/image analyses of subjects could incorporate physiological and/or subconscious responses into the study of consumers. There is a growing interest in understanding the role of physiological reactions of participants toward the sensory assessments of foods.

Furthermore, sensory laboratories use isolated booth environments that are designed to control against the effects of non-product factors such as the external aromas, light distractions, and noises of various surrounding environments. However, some researchers argue that this setting does not represent the actual conditions in which consumers taste their products. Highly controlled testing conditions may lack ecological validity that can lead to a biased evaluation of the sensory attributes by consumers. Recently, the use of virtual reality (VR) environments has become popular for testing the effects of context on the sensory experience. Sensory analysis of foods, beverages, and packaging are incorporating quantitative aspects of culture, behavior/mood, and environment, which require novel approaches.

This book, reprinted from articles published in the Special Issue "Novel Techniques to Measure the Sensory, Emotional, and Physiological (Biometric) Responses of Consumers toward Foods and Packaging" of the journal *Foods*, aims to provide a deeper understanding of novel techniques to measure the different sensory, emotional, and physiological responses toward different products.

Damir Torrico
Editor

Editorial

Novel Techniques to Measure the Sensory, Emotional, and Physiological Responses of Consumers toward Foods

Damir D. Torrico

Department of Wine, Food and Molecular Biosciences, Faculty of Agriculture and Life Sciences, Lincoln University, Lincoln 7647, New Zealand; Damir.Torrico@lincoln.ac.nz; Tel.: +64-3-423-0641

Citation: Torrico, D.D. Novel Techniques to Measure the Sensory, Emotional, and Physiological Responses of Consumers toward Foods. *Foods* **2021**, *10*, 2620. https://doi.org/10.3390/foods10112620

Received: 21 October 2021
Accepted: 25 October 2021
Published: 29 October 2021

Publisher's Note: MDPI stays neutral with regard to jurisdictional claims in published maps and institutional affiliations.

Sensory science is an evolving field that has been incorporating technologies from different disciplines. Sensory evaluation is very important for the food and beverage industries as it can provide applied insights regarding the reactions of consumers towards products. Traditional techniques such as discrimination, descriptive and affective tests have been utilized for understanding the sensory properties of foods and evaluating the hedonic and emotional responses of consumers. However, these techniques have their drawbacks and sensory scientists have been working on improving these methodologies or developing new measurement systems. For instance, sensory scientists have been investigating alternative measurements of consumers by using physiological responses. On the other hand, research has been conducted to understand the effects of context in the consumer evaluation of foods and beverages. Moreover, techniques such as home-use tests or virtual reality evaluations are being regarded as having more ecological validity since they require participants to taste the products in their day-to-day environments. This Special Issue of *Foods* aimed to understand the development of novel techniques to measure the sensory, emotional, and physiological responses of consumers toward food.

Traditional sensory methods use self-reported responses to gather information from trained panels or untrained consumers. However, these techniques have the limitation of not capturing the totality of responses from these participants [1]. Moreover, self-reported responses in sensory tests are exposed to cognitive biases since they are not generated immediately during decision making [2]. A complementary measurement of self-reported responses is the incorporation of physiological reactions that can be achieved by the use of biometrics. Responses such as heart rate, body temperature, and facial expressions have been shown to tap into the unconscious responses of consumers in sensory tests. In a study that evaluated energy drinks, Mehta, Sharma, Kanala, Thakur, Harrison and Torrico [1] investigated the self-reported responses (including hedonic and emotional) and facial expressions to determine the degree to which participants liked these products. They concluded that participants elicited more self-reported positive than negative emotions for the energy drinks, and that the implicit emotional responses through facial expressions were important in the discrimination of the products. In a study of emotions toward beer products, Gonzalez Viejo et al. [3] concluded that sweet beers were associated with higher liking and positive emotions; on the other hand, bitter beers were associated with negative self-reported emotions and negative feelings expressed through facial expressions and other physiological measurements.

These previous studies showed the importance of assessing biometric responses when testing for familiar products. However, consumers nowadays are exposed to a wide variety of foods from different countries and production systems. An important food trend that is influencing consumers' purchasing decisions is the sustainability of foods/beverages production systems. The alternative protein movement has been pushing food developers to reformulate existing products with new ingredients and at the same time to keep the acceptability of those products at similar or higher levels. In this context, understanding the consumer assessments of emerging ingredients such as plant- and insect-based products is becoming crucial for companies that are implementing alternative protein foods. In this regard, Gupta et al. [4] evaluated plant-based yogurt products using hedonic, emotional (check-all-that-apply methodology with emotional terms and emojis), and physiological responses of consumers. They found that the use of emojis was effective in the characterization of cross-cultural preferences of different yogurt formulations compared to those of hedonic ratings and facial expressions. In another study, Fuentes, Wong and Gonzalez Viejo [2] evaluated the sensory and biometric responses of consumers toward insect-based food snacks. They concluded that snacks containing visible insects in their presentations had a lower degree of acceptance and negative emotional responses compared to snacks containing non-visible insects or no insects in their formulations. This study demonstrated that the appearance of products could have a profound effect on the hedonic and emotional assessments of foods containing insects.

Consumers' purchasing decisions are not only affected by the intrinsic sensory properties of the products but also by the extrinsic information that is provided at the time of purchasing. De Wijk et al. [5] evaluated the effects of familiarity and branding on the sensory experiences of soy sauces using facial expressions and video-based heart rate measurements. They found that liking and arousal toward the products were affected by taste but not by branding and familiarity. On the other hand, biometric measurements (facial expressions and heart rate) were affected by the branding and familiarity of the soy sauces. This study also showed that the assessment of facial expressions and heart rate can be done remotely without the use of centrally located laboratories, and only by using image and video recording devices. Another important parameter to measure the implicit (unconscious) reactions of consumers is the Approach Avoidance Task (AAT). Humans, in general, tend to avoid (leaning backwards) the stimuli that they dislike, and they tend to approach (leaning forward) the stimuli that they like when they are doing a sensory test. Brouwer et al. [6] evaluated the AAT using a mobile version that consumers can use at their homes using their phones to measure their reactions (AAT and valance ratings) toward pictures of palatable and unpalatable foods. They found that the ATT measurements obtained by this mobile version could complement the rating scores of participants, which can be useful in understanding the intrinsic reactions of consumers.

Traditional consumer tests focused on understanding the links between foods and consumers. Measuring the hedonic responses and preferences is very important for companies when launching new products into the marketplace. Traditional consumer tests use booths to isolate participants from external factors (noises, smells). This is a common practice in sensory evaluation since consumers can be easily distracted from the sensory assessment and their responses can be biased by these external factors. However, isolated booths might not represent the "real" environments where consumers usually taste their products. In some cases, the combinations of environment and product can enhance the sensory experiences of consumers. For that reason, new research has been done to understand the effect of environments on sensory perception. Montero et al. [7] studied the consumer acceptance of ready-to-eat (RTE) meals during storage using a home-use test (HUT). They concluded that the HUT methodology can enhance the sensory experience of consumers when tasting the products.

In some cases, HUT might not be possible to conduct due to budget constraints or the requirement of specific preparations of the products. For those factors, researchers have started to use alternative methods to measure the effect of the environment on the consumer experience. New technologies such as virtual and augmented reality are becoming very useful when testing for the environment. Crofton et al. [8] explored the effects of immersive virtual reality (VR) technologies on the sensory perception of beef and chocolate. They found that VR had a significant effect on participants' hedonic responses to the products and that consumers' level of engagement was different depending on the context. In a similar study, Kong et al. [9] demonstrated that VR had slight effects on the emotions of consumers towards chocolate products. In a study of wines, Torrico et al. [10] found that VR environments, in general, affected the perception of floral aroma and emotions.

In conclusion, this Special Issue showed that sensory evaluation is changing from the sole use of rating/ranking scales to the incorporation of biometric measurements and the use of immersive technologies. The development of new sensory techniques is an ongoing task, which is constantly looking for accurate measurements of consumers' responses. The editor hopes that the findings from this Special Issue can help the broader scientific community to understand the use of novel sensory science techniques that can be used in the evaluation of food and beverage products.

Funding: This research received no external funding.

Conflicts of Interest: The authors declare no conflict of interest.

References

1. Mehta, A.; Sharma, C.; Kanala, M.; Thakur, M.; Harrison, R.; Torrico, D.D. Self-reported emotions and facial expressions on consumer acceptability: A study using energy drinks. *Foods* **2021**, *10*, 330. [CrossRef] [PubMed]
2. Fuentes, S.; Wong, Y.Y.; Gonzalez Viejo, C. Non-invasive biometrics and machine learning modeling to obtain sensory and emotional responses from panelists during entomophagy. *Foods* **2020**, *9*, 903. [CrossRef] [PubMed]
3. Gonzalez Viejo, C.; Villarreal-Lara, R.; Torrico, D.D.; Rodríguez-Velazco, Y.G.; Escobedo-Avellaneda, Z.; Ramos-Parra, P.A.; Mandal, R.; Pratap Singh, A.; Hernández-Brenes, C.; Fuentes, S. Beer and consumer response using biometrics: Associations assessment of beer compounds and elicited emotions. *Foods* **2020**, *9*, 821. [CrossRef] [PubMed]
4. Gupta, M.; Torrico, D.D.; Hepworth, G.; Gras, S.L.; Ong, L.; Cottrell, J.J.; Dunshea, F.R. Differences in Hedonic Responses, Facial Expressions and Self-Reported Emotions of Consumers Using Commercial Yogurts: A Cross-Cultural Study. *Foods* **2021**, *10*, 1237. [CrossRef] [PubMed]
5. De Wijk, R.A.; Ushiama, S.; Ummels, M.; Zimmerman, P.; Kaneko, D.; Vingerhoeds, M.H. Reading Food Experiences from the Face: Effects of Familiarity and Branding of Soy Sauce on Facial Expressions and Video-Based RPPG Heart Rate. *Foods* **2021**, *10*, 1345. [CrossRef] [PubMed]
6. Brouwer, A.-M.; van Beers, J.J.; Sabu, P.; Stuldreher, I.V.; Zech, H.G.; Kaneko, D. Measuring Implicit Approach–Avoidance Tendencies towards Food Using a Mobile Phone Outside the Lab. *Foods* **2021**, *10*, 1440. [CrossRef] [PubMed]
7. Montero, M.L.; Garrido, D.; Gallardo, R.K.; Tang, J.; Ross, C.F. Consumer Acceptance of a Ready-to-Eat Meal during Storage as Evaluated with a Home-Use Test. *Foods* **2021**, *10*, 1623. [CrossRef] [PubMed]
8. Crofton, E.; Murray, N.; Botinestean, C. Exploring the effects of immersive virtual reality environments on sensory perception of beef steaks and chocolate. *Foods* **2021**, *10*, 1154. [CrossRef] [PubMed]
9. Kong, Y.; Sharma, C.; Kanala, M.; Thakur, M.; Li, L.; Xu, D.; Harrison, R.; Torrico, D.D. Virtual reality and immersive environments on sensory perception of chocolate products: A preliminary study. *Foods* **2020**, *9*, 515. [CrossRef] [PubMed]
10. Torrico, D.D.; Han, Y.; Sharma, C.; Fuentes, S.; Gonzalez Viejo, C.; Dunshea, F.R. Effects of context and virtual reality environments on the wine tasting experience, acceptability, and emotional responses of consumers. *Foods* **2020**, *9*, 191. [CrossRef] [PubMed]

Article

Self-Reported Emotions and Facial Expressions on Consumer Acceptability: A Study Using Energy Drinks

Annu Mehta, Chetan Sharma, Madhuri Kanala, Mishika Thakur, Roland Harrison and Damir Dennis Torrico *

Department of Wine, Food and Molecular Biosciences, Lincoln University, Lincoln 7647, New Zealand; Annu.Mehta@lincolnuni.ac.nz (A.M.); chetan.sharma@lincoln.ac.nz (C.S.); Madhuri.KanalaManjunath@lincolnuni.ac.nz (M.K.); Mishika.Thakur@lincolnuni.ac.nz (M.T.); Roland.Harrison@lincoln.ac.nz (R.H.)
* Correspondence: Damir.Torrico@lincoln.ac.nz; Tel.: +64-3-423-0641

Abstract: Emotional responses elicited by foods are of great interest for new product developers and marketing professionals, as consumer acceptance proved to be linked to the emotions generated by the product in the consumers. An emotional measurement is generally considered an appropriate tool to differentiate between the products of similar nutritional value, flavour, liking and packaging. Novel methods used to measure emotions include self-reporting verbal and visual measurements, and facial expression techniques. This study aimed to evaluate the explicit and implicit emotional response elicited during the tasting of two different brands (A and B) of energy drinks. The explicit response of consumers was assessed using liking (nine-point hedonic scale), and emotions (EsSense Profile®—Check-All-That-Apply questionnaire), and implicit emotional responses were evaluated by studying facial expressions using the Affectiva Affdex® software. The familiarity of the product and purchase intent were also assessed during the study. The hedonic rating shows a significant difference in liking between the two brands of energy drink during the tasting session. For the explicit emotional responses, participants elicited more positive emotions than the negative emotions for both energy drinks. However, participants expressed "happy", "active" and "eager" emotions more frequently for energy drink A. On the other hand, the implicit emotional responses through facial expressions indicated a high level of involvement of the participants with energy drink B as compared to energy drink A. The study showed that overall liking and the explicit and implicit emotional measurements are weakly to moderately correlated.

Keywords: emotions; EsSense profile®; facial expressions; purchase intention; energy drinks

Citation: Mehta, A.; Sharma, C.; Kanala, M.; Thakur, M.; Harrison, R.; Torrico, D.D. Self-Reported Emotions and Facial Expressions on Consumer Acceptability: A Study Using Energy Drinks. *Foods* **2021**, *10*, 330. https://doi.org/10.3390/foods 10020330

Academic Editor: Derek V. Byrne
Received: 2 October 2020
Accepted: 29 January 2021
Published: 4 February 2021

Publisher's Note: MDPI stays neutral with regard to jurisdictional claims in published maps and institutional affiliations.

1. Introduction

Globalization, which involves a fierce competition among companies to retain consumers and simultaneously to find new ones, is the driving force that keeps food and other industries innovating to make well-informed decisions in the marketplace. In the course of innovation, many concepts are brainstormed, and only a few selective ideas are taken forward for bench- or pilot-scale testing. However, the success of these prototypes in the marketplace is not guaranteed. Indeed, 50–70% of newly launched food products do not last long in the market [1], despite their intensive market research. Sensory and consumer sciences provide a few tools in this context [2–5] to better understand products [6–9] and population categories [10], and to minimize the risk of failure [10,11]. One of the most popular and extensively used methods to quantify affective responses in sensory science is the acceptability test using the nine-point hedonic scale [12]. However, product sensory liking does not necessarily always convert into a purchase [13], and formulators should look beyond the liking scores to make a product successful [14]. Thus, many other methods [3,13,15] are being explored to provide insights into food-choice behaviours. A novel sensory and emotional approach is the use of non-verbal cues, such as facial expressions, which can communicate highly detailed information about the individual's experiences [16] and helps in understanding product liking and different emotions that influence the purchase intention. Emotions play a significant role in the comprehension of food preferences and consumers' likings. Moreover, emotions are decisive factors in our food choices since the consumed foods can evoke certain emotions [14,17,18]. Several factors affect emotions, such as age, satiety, health, economic condition, and expectations. In addition to these, two types of emotional responses, conscious and unconscious, can be elicited by consumers when exposed to different products [19,20]. However, most market research is based on conscious arousal and measured with self-reporting scales [21,22], such as EsSense Profile® [23] and PrEmo® [24].

Recent studies using the EsSense profile® questionnaire have validated its discriminating power within and between food product categories [25]. The EsSense Profile® questionnaire is cost-effective, easy to use and interpret, covers a wide range of emotions and has provided rich insights into the consumers' perceptions as well as the liking of products such as beer [26], wine [27,28] and coffee [29,30]. However, the explicit method of emotional measurement requires cognitive thinking to convert the experiences into expressions, which sometimes lose the actual meaning of the emotions felt.

Automatic facial expression recognition (hereafter AFER) is one of the important novel methods to study emotional responses and human behaviours. AFER is a non-verbal and arguably universal language [31] that helps communicate countless emotions, such as happiness, sadness, anger, fear, surprise, and others among humans. Recent studies have been conducted to find the correlation between the AFER and emotions, and how both are comparatively linked with the self-reported likings [32]. Infants and children have been using facial expressions to communicate their emotions and feelings [33], especially in the case of sweet foods, when they smack their lips, protrude the tongue and smile; while, for bitter food, they wrinkle their nose and turn their heads [34]. Researchers have found that facial expressions can recognize and differentiate the basic tastes and odours [35] and found that consumers elicit negative expressions (dislike) more accurately than positive expressions [36]. AFER and autonomous nervous system responses provide insights into food preference in relation to food properties for different kind of foods [37–39] and its influence on the purchase behaviours of consumers. The facial expressions can elucidate the consumer acceptability of products based on emotional responses and familiarity. In the present study, energy drinks were selected as the beverage model for evaluating the hedonic and emotional responses of consumers. Energy drinks are soft beverages that contain several stimulants such as caffeine and glucose and have a wide range of flavours and mouthfeels [40], which can affect mood and mental energy [41]. Many studies have proved the effect of caffeine or the combination of caffeine with other ingredients on the mood and cognitive performance [41]. Caffeine in doses of 75 and 150 mg enhances positive emotions (such as happiness and calmness), while it reduces tenseness [42,43] and caffeinated taurine drinks improve alertness [44] and attention [45]. The overall liking and explicit emotion measurements were investigated using a self-reported questionnaire, and implicit emotion measurements were observed from the facial expressions using the Affectiva® software. The study aimed to investigate whether self-reporting liking and explicit emotion measurements provide similar differentiation among the samples. It also evaluated whether the positive/negative emotions elicited during the implicit emotion measurements were correlated to the sensory attributes of the products. The explicit and implicit measurements of emotions on the differentiation of samples were also investigated. Finally, self-reported liking, and explicit and implicit measurements of emotions were studied on the samples' differentiation. Therefore, the following hypotheses were proposed for this research: H1—The self-reported liking and positive explicit emotion measurement are positively correlated; H2—The positive facial expression emotions and liking will depict similar product differentiation; and H3—The self-reported liking, explicit and implicit emotions, exhibit similar behaviours.

2. Material and Methods

2.1. Participants

Forty-seven participants (male/female 21/26), who were 20–40 years of age, were recruited via email from Lincoln University for the research experiment. The majority of the participants were students of Asian origin (India (N = 27), China (N = 13), Vietnam (N = 3), Korea (N = 1), Hispanic (N = 2) and Cambodia (N = 1)). The participants were students of Lincoln University and were of Asian origin. Facial expressions vary with different cultures and ethnicities [46,47]; therefore, ethnic-specific facial expressions data were required to understand the implicit emotions depicted by consumers after tasting energy drinks. The participants provided written consent for any sensory deficiency such as anosmia and ageusia, a tasting and video recording session as per ethical requirements—Human Ethics (approval: 2019-68). The selected participants (who self-reportedly did not have any of the previously described sensory deficiencies) were not trained for the experiment, and no prior information regarding the study was disclosed to them. The panellists had previously participated in other focus group studies related to other food products such as chocolates and wines and had experience with this kind of study. The criteria to select panellists were that they should be familiar with the product and consume energy drink at least once a month. The general instructions regarding the procedure were given to the participants, providing information regarding the video recording of their tastings. Participants were asked to look at the camera and focus on evaluating the sensory characteristics of the products. The study was conducted in a sensory laboratory of the Department of Wine, Food and Molecular Biosciences, Lincoln University, New Zealand, which meets the sensory evaluation requirements listed in ISO 6658, 2005 and GB 13868, 2009. In regard to the consumer sample size used, a power analysis to test how this experiment performed was conducted. With a difference in means of 0.81 for overall liking, the power of this experiment was ~0.7; therefore, the probability of Type II error in this experiment is medium to low (~0.3) for this type of consumer's assessments [48]. In addition, based on an extensive study of acceptability tests, Gacula Jr and Rutenbeck [49] estimated that the correct sample size for the consumer's evaluations was between 40 and 100 consumers. However, increasing the number of participants can help to minimize the Type II error, increasing the power of the experiment. The samples (~10 mL) were stored and served at a refrigerated temperature of 4 °C in transparent plastic cups marked with three-digit random codes in a white tray. Crackers (Arnotts, Australia) and water were served to rinse the palate after each sample and were asked to have a five-minute break before the next sample to avoid sensory fatigue.

2.2. Sample Selection

As a preliminary test, a focused group of four trained panellists evaluated five different brands of energy drinks ((Red Bull, Red Bull GmbH, Salzburg, Austria), (Monster, Monster Beverage Corporation, Corona, CA, USA), (Mother, Monster Beverage Corporation, Corona, CA, USA), (V Guarana, Frucor, Auckland, New Zealand), and (Rockstar, Rockstar, Inc., Las Vegas, NV, USA)) for taste and flavour acceptability. After discussion in a focus group (N = 4) and the descriptive analysis of the focus group results, two energy drinks with the highest and the lowest acceptability scores were selected to delineate the difference between the products. These liking differences were used for polarizing the facial emotions that consumers might express during the tasting The Rockstar energy drink manufactured by Rockstar, Inc, The United States (Energy drink A) (*ingredients*: carbonated water, sucrose, glucose, citric acid, taurine, natural and artificial flavours, sodium citrate and caffeine, benzoic acid, caramel colour, sorbic acid, L-cartinine, inositol, niacinamide, calcium pantothenate, milk thistle extract, gingko, biloba leaf extract, guarana seed extract, panax. ginseng root extract, riboflavin, pyridoxine hydrochloride, cyanocobalamin) and V energy drink produced at Frucor, New Zealand (Energy drink B) (*ingredients*: carbonated water, sugar, acidity regulator (citric acid and sodium citrate), taurine, guarana extract (0.12%), colour (caramel), glucuronolactone, caffeine, inositol, vitamins (niacin (B3), pantothenic acid, B6, riboflabin B2, (B12), flavours and contains wheat derivatives) were selected for the experiment.

2.3. Traditional Technique of Acceptance Measurement

The traditional method of acceptability measurement is to ask participants for attributes and overall liking using a 9-point hedonic scale (from 1 = extremely dislike, to 9 = extremely like) [12]. In other words, appearance liking, aroma liking, flavour liking, sweetness liking, bitterness liking, aftertaste liking, mouthfeel liking and overall liking were used for evaluating both energy drinks.

2.4. Familiarity and Purchase Intent

The familiarity of the energy drink was assessed using a 5-point scale from 1 (not familiar) to 5 (very familiar). Many studies related to children or adults having proved that in the presence of a familiar person or a product, facial expressions are affected and can cause familiarity bias [50]. Therefore, familiarity was taken as a covariable of the liking experiment in this study. For purchase intent, the binomial scale (Yes or No) was used to answer a question *"Will you purchase this product in the future based on the taste/flavour characteristics?"*

2.5. Self-Reported Emotions by EsSense® Profile

A total of 21 emotions from EsSense Profile®, such as "active", "adventurous", "bored", "daring", "disgusting", "eager", "energetic", "good", "happy", "interested", "joyful", "mild", "pleasant", "satisfied", "warm", "wild", "anger", "sadness", "surprised", "fear" and "contempt" were used for this study. The aforementioned emotion terms were selected after consensus by a focus group ($N = 4$). The emotional terms were selected on the basis of the frequency of use (>20%) categorization and in relation to the food tested. The check-all-that-apply (CATA) methodology was used for the consumer study to evaluate the emotions elicited by the energy drinks.

2.6. Implicit Emotions by Automated Facial Expression Response Measurement

The facial expressions of 30 among 47 panellists were evaluated using Affdex (Affectiva Inc., Waltham, MA, USA) based on the facial inputs. The automated facial coding engine (AFFDEX) was integrated with iMotions Facial Expression Analysis Module (iMotions, Inc., Boston, MA, USA) for decoding the facial emotions using a group of action units (Table 1). The iMotions Facial Expression Analysis Module detects and extracts seven core emotions (joy, anger, fear, disgust, contempt, sadness, and surprise) (shown in Figure 1) and 20 facial expression measures (action units). The action units describe the movements of facial muscles. Emotions are displayed by the movements of a certain number of combined facial muscles. iMotions module also provides timelines annotations, and scores of engagement and valence to provide insight into the facial emotions (https://imotions.com/). The intensity for emotional expression varies from 0 (no expression) to 100 (expression present). The facial expression data collected was from the first three seconds after the participants put the energy drink in their mouths. This is based on previous studies in drinks, in which it was demonstrated that automatic nervous system (AND) responses are expressed immediately after participants are exposed to the stimuli [37,51].

Table 1. Action units of the facial muscle movements expressing emotions in Affective® https://imotions.com/blog/facial-action-coding-system/.

Emotion	Action Units	Description
Happiness/joy	6 + 12	Cheek raiser, lip corner puller
Sadness	1 + 4 + 15	Inner brow raiser, brow lowerer, lip corner depressor
Surprise	1 + 2 + 5 + 26	Inner brow raiser, outer brow raiser, upper lid raiser, jaw drop
Fear	1 + 2 +4 + 5 + 7 + 20 + 26	Inner brow raiser, outer brow raiser, brow lowerer, upper lid raiser, lid tightener, lip stretcher, jaw drop
Anger	4 + 5 + 7 + 23	Brow lowerer, upper lid raiser, lid tightener, lip tightener
Disgust	9 + 15 + 16	Nose wrinkler, lip corner depressor, lower lip depressor
Contempt	12 + 14 (on one side of the face)	Lip corner puller, dimpler

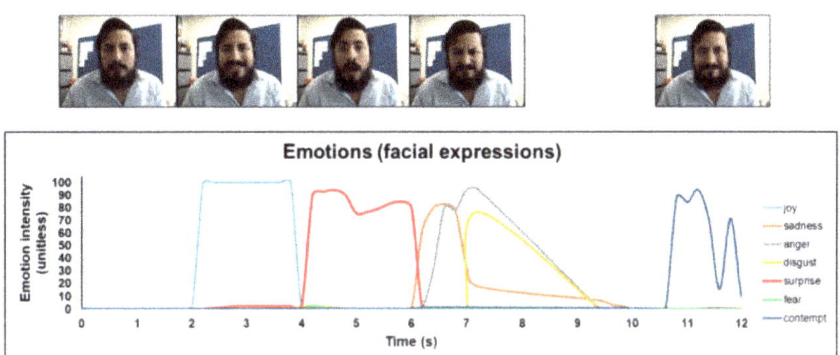

Figure 1. Different emotions: sadness, happy (joy), surprise, disgust, anger, fear, and contempt elicited by a participant. This figure shows 12 s of recording as a demonstration of the different faces' movements and their relationship with emotions. Actual facial expressions were taken during three seconds after swallowing the sample.

2.7. Testing Procedure

The experimental details were explained to participants prior to the commencement of the experiment. The participants were asked to drink 10 mL of the sample in one swallow [52,53], and wait for 15 s, looking straight to the tablet for the recording of the facial expressions. After 15 s, the participants were allowed to fill out the questionnaire section regarding liking and explicit emotions based on a previous study [37]. Therefore, the chances of cognitive bias were minimized. The participants were asked to taste the samples that were served in a 30 mL plastic cups maintained at a temperature of 4 °C and answer the questions related to the sensory attributes, liking, familiarity, purchase intent and emotions felt during the tasting using the questionnaire generated by the RedJade® Sensory Software (RedJade®, Martinez, CA, USA). The presentation order of the samples was randomized. The participants tasted the samples under blind conditions; however, facial expressions were recorded for 3 s immediately after the first sip of the sample. The facial expressions of participants were recorded using a camera with a video resolution of 4K (UHD) at 30 FPS (X450, Kaiser Bass, Australia) adjusted in front of participants for the recording of the facial reactions. As a token of appreciation, a can of energy drink was gifted to participants after the experiment.

2.8. Statistical Analysis

The hedonic scores of sensory attributes and the liking of the energy drinks during the tasting session were analysed using an analysis of variance (ANOVA). A Tukey test was used as a post hoc data analysis technique to determine the differences between the samples. The significance level (α) was set at 5%. To analyse the relationship between energy drinks and familiarity, the chi-square test for homogeneity was performed. EsSense Profile® data were analysed using the XLSTAT Statistical Software (XLSTAT Version 2019.4.2, Addinsoft, New York, NY, USA). The frequency counts of 21 emotion words which described the samples were calculated. Cochran's Q test was used to find the difference between the samples by evaluating each emotion word used in the CATA questionnaire. The values of purchase intent were statistically analysed for multiple comparisons using the Cochran's Q test. Facial expression data were collected through the Affectiva Affdex (Affectiva Inc., Waltham, MA, USA) software. The analysis of variance (ANOVA) technique, through Minitab® 18 (Version 10.0.17763 Build 17763, State College, PA, USA), was used to locate significant differences among emotions. Pearson correlation coefficients (r) among the sensory attributes, familiarity, and facial expressions were calculated and plotted using a correlation matrix. Multivariate analysis of variance (MANOVA) was used to determine the significant difference among facial expressions and correlations among all sensory attributes were tested. Based on the MANOVA results, a principal components analysis (PCA) bi-plot was made. The relationship between each emotion method using overall liking as the response variable was analysed using a linear regression model.

3. Results

3.1. Traditional Technique of Acceptance Measurement, Familiarity and Purchase Intent

The mean and standard deviation of sensory attributes, overall liking and familiarity of the energy drinks A and B are shown in Table 2. No significant differences ($p > 0.05$) were found between both the energy drinks for all the sensory attributes (appearance, aroma, taste/flavour, sweetness, bitterness, mouthfeel and aftertaste). There was a significant difference in the overall liking of the energy drinks. Energy drink A had a significantly higher overall liking score (6.79) compared to that of energy drink B (5.98). Based on the chi-square test $X2$ (4, $N = 94$) = 5.25, $p = 0.26$, no significant difference was found among the samples. The expected value of the last category of the attribute was more than the observed value; therefore, the result's interpretation was sceptical.

Table 2. Sensory attributes, overall liking and familiarity of energy drink A and B after tasting.

Attributes	Energy Drink A	Energy Drink B	F Value *	Pr > F *
Appearance	6.96 ± 1.33 [a]	6.53 ± 1.38 [a]	2.31	0.13
Aroma	6.89 ± 1.70 [a]	6.40 ± 1.51 [a]	2.18	0.14
Flavour	6.72 ± 1.75 [a]	6.02 ± 1.94 [a]	3.39	0.07
Sweetness	6.55 ± 1.82 [a]	6.23 ± 1.90 [a]	0.69	0.41
Bitterness	5.75 ± 1.93 [a]	5.32 ± 1.82 [a]	1.21	0.27
Mouthfeel	6.75 ± 1.91 [a]	6.15 ± 2.03 [a]	2.15	0.15
Aftertaste	6.28 ± 2.03 [a]	5.51 ± 2.17 [a]	3.13	0.08
Overall liking	*6.79 ± 1.67* [a]	*5.98 ± 2.03* [b]	*4.46*	*0.04*
Familiarity **	2.72 ± 1.30 [a]	2.43 ± 1.02 [a]	$X2 = 5.25$	$p = 0.26$

Sensory attributes and overall liking of energy drink A and energy drink B were measured by 9-point hedonic scale (1 = extremely disliked and 9 = extremely liked) and familiarity between the energy drinks were assessed with 5-point categorical scale (1 = not at all familiar and 5 = extremely familiar). Bold italicized values indicate that the parameter was significantly different from 0 ($p < 0.05$). * F value = mean square or mean square error. Effects were considered significant if Pr (probability) > F was <0.05 (bold probability and F value). ** Familiarity was analysed by chi-square at a 95% confidence level. [a,b] Means with different superscripts in each row indicate significant differences ($p < 0.05$).

The frequency at which a consumer's intent to buy energy drinks based on the sensory attributes are shown in Table 3. A total of 68.09% of participants intended to buy energy drink A, while 55.32% of participants expected to buy energy drink B. No significant differences in the purchase intent of energy drink A and B were found.

Table 3. Purchase intent frequencies of energy drinks during the tasting session.

Energy Drinks	Willingness to Purchase (%)
A	68.09 [a]
B	55.32 [a]

Cochran's Q is used to find the difference between energy drinks. The table shows a percentage of consumers willing to buy the energy drink after the tasting session of the energy drinks. Different superscript letters in each row indicate significant differences ($p < 0.05$).

3.2. Self-Reported Emotion Measurements

The frequencies of self-reported emotions data are shown in Table 4. The significant differences in the emotion terms "active" and "interested" were reported in the energy drinks during the tasting. The selection frequency of "active" for energy drink A was 49%, while the selection frequency of "active" for energy drink B was 34%. The selection frequency of "interested" for energy drink A (32%) was almost double than the selection frequency of "interested" for energy drink B (13%). No significant differences were found in other reported emotions for both energy drinks during the tasting. Overall, the participants felt more positive emotions ("good", "happy", "interested", "joyful" and "pleasant") for energy drink A than that for energy drink B. None of the participants felt "sadness" while tasting sample A or B.

Table 4. Emotions felt after tasting energy drink A and B from Cochran's Q test.

Attributes	Energy Drink A	Energy Drink B
Active	***0.49*** [a]	***0.34*** [b]
Adventurous	0.19 [a]	0.13 [a]
Bored	0.11 [a]	0.04 [a]
Daring	0.06 [a]	0.06 [a]
Disgusted	0.06 [a]	0.11 [a]
Eager	0.11 [a]	0.13 [a]
Energetic	0.36 [a]	0.43 [a]
Good	0.49 [a]	0.40 [a]
Happy	0.28 [a]	0.15 [a]
Interested	***0.32*** [a]	***0.13*** [b]
Joyful	0.30 [a]	0.15 [a]
Mild	0.26 [a]	0.19 [a]
Pleasant	0.34 [a]	0.28 [a]
Satisfied	0.26 [a]	0.26 [a]
Warm	0.15 [a]	0.09 [a]
Wild	0.09 [a]	0.09 [a]
Anger	0.02 [a]	0.02 [a]
Sadness	0.00 [a]	0.00 [a]
Surprised	0.11 [a]	0.23 [a]
Fear	0.04 [a]	0.02 [a]
Contempt	0.11 [a]	0.09 [a]

A check-all-that-apply (CATA) questionnaire was used to select emotions related to the sample, and Cochran's Q is used to find the difference between the products. Bold italicized values indicate that the parameter was significantly different ($p < 0.05$). [a,b] Means with different superscripts in each row indicate significant differences ($p < 0.05$).

3.3. Automated Facial Expression Response Measurements

The mean and standard deviation for the facial expression parameters (joy, sadness, disgust, contempt, anger, fear, surprise, valence, engagement and smile) are shown in Table 5. No significant differences in the facial expressions were reported for both products. Some marginal differences were observed in a few emotion categories, such as "smile", "engagement", "joy", "disgust", and "surprise". The intensity of "engagement" was highest for both sample types (2.79 and 1.25 for sample A and B, respectively), followed by "contempt" (2.36 and 0.15 for sample A and B, respectively) and "joy" (0.56 for sample A and 0.97 for sample B). For energy drink B, the participants elicited slightly more "joy" and "smile" compared to that of energy drink A. The intensities of "joy" and "smile" were almost double in energy drink B than that of in energy drink A. The negative emotions such as "sadness", "anger", and "fear" were marginally elicited in both sample types.

Table 5. Facial expressions and facial features during the tasting of energy drinks A and B.

Parameters	Energy Drink A	Energy Drink B	F Value	Pr > F *
Engagement	2.79 ± 8.27 [a]	1.25 ± 5.11 [a]	0.54	0.47
Contempt	2.36 ± 10.07 [a]	0.15 ± 0.07 [a]	1.25	0.27
Smile	0.83 ± 3.07 [a]	1.24 ± 5.11 [a]	0.1	0.75
Joy	0.56 ± 2.55 [a]	0.97 ± 4.87 [a]	0.12	0.73
Valence	0.44 ± 2.51 [a]	0.52 ± 4.63 [a]	0.01	0.95
Disgust	0.36 ± 0.22 [a]	0.78 ± 2.10 [a]	0.86	0.36
Surprise	0.30 ± 0.43 [a]	0.17 ± 0.08 [a]	2.44	0.13
Sadness	0.02 ± 0.01 [a]	0.02 ± 0.01 [a]	0.001	0.98
Anger	0.02 ± 0.00 [a]	0.02 ± 0.00 [a]	0.16	0.69
Fear	0.00 ± 0.00 [a]	0.01 ± 0.015 [a]	1.13	0.29
Attention	67.90 ± 29.15 [a]	62.49 ± 30.10 [a]	0.39	0.54
Eye closure	28.12 ± 41.23 [a]	18.96 ± 35.57 [a]	0.67	0.42
Mouth open	2.43 ± 10.18 [a]	0.58 ± 1.39 [a]	0.84	0.36
Dimpler	2.38 ± 7.72 [a]	0.15 ± 0.44 [a]	2.19	0.15
Smirk	2.12 ± 9.00 [a]	0.13 ± 0.54 [a]	1.27	0.27
Brow raise	2.03 ± 6.33 [a]	0.36 ± 0.85 [a]	1.79	0.19
Lip suck	1.81 ± 5.44 [a]	1.91 ± 8.75 [a]	0.00	0.97
Inner brow raises	0.72 ± 1.86 [a]	0.71 ± 2.76 [a]	0.00	0.99
Lip press	0.41 ± 1.20 [a]	0.08 ± 0.16 [a]	1.88	0.18
Jaw drop	0.40 ± 0.68 [a]	0.21 ± 0.25 [a]	1.77	0.19
Chin raise	0.27 ± 0.69 [a]	0.93 ± 3.79 [a]	0.63	0.43
Lid tighten	0.23 ± 0.47 [a]	0.31 ± 0.69 [a]	0.17	0.68
Nose wrinkle	0.16 ± 0.41 [a]	2.24 ± 7.55 [a]	1.59	0.21
Brow furrow	0.09 ± 0.30 [a]	0.21 ± 0.79 [a]	0.38	0.54
Lip pucker	0.08 ± 0.17 [a]	0.09 ± 0.18 [a]	0.03	0.87
Cheek raise	0.04 ± 0.14 [a]	0.02 ± 0.04 [a]	0.26	0.61
Lip stretch	0.04 ± 0.14 [a]	1.11 ± 5.63 [a]	0.76	0.39
Eye widen	0.02 ± 0.04 [a]	1.36 ± 5.35 [a]	1.31	0.26
Upper lip raise	0.02 ± 0.07 [a]	0.31 ± 1.06 [a]	1.63	0.21
Lip corner depressor	0.01 ± 0.02 [a]	0.01 ± 0.02 [a]	0.03	0.86

Means values of the different facial expressions with different superscripts in each row indicate a significant difference ($p < 0.05$). * F value = Mean square or mean square error. Effects were considered significant if Pr (probability) > F was <0.05 (bold probability and F value). [a] Means with different superscripts in each row indicate significant differences ($p < 0.05$).

3.4. Correlation and Multivariate Analysis of Hedonic and Facial Expression Responses

The Pearson correlation coefficient matrix (Table 6) shows the correlations between various variables (sensory attributes, overall liking, familiarity and AFER) for energy drinks A and B. Familiarity was positively correlated with overall liking ($r = 0.44$) and various sensory attributes such as the liking of taste/flavour ($r = 0.40$), sweetness ($r = 0.38$), bitterness ($r = 0.44$), mouthfeel ($r = 0.40$) and aftertaste ($r = 0.46$). The appearance of the product was negatively correlated to the facial expression of "sadness" ($r = -0.33$). The facial expression "anger" was negatively correlated with the liking of sensory attributes such as sweetness ($r = -0.38$) and aftertaste ($r = -0.29$). The results of the principal components analysis (PCA), which explains the relationship between emotions and hedonic attributes are shown in Figure 2. The principal component one (PC1) explained 31.59% of total inertia, while principal component two (PC2) explained 12.68%, respectively. The overall liking vector was positively related to familiarity, sensory attributes (taste/flavour, mouthfeel, aroma, sweetness, bitterness and aftertaste) and emotions ("disgust" and "sadness"), and negatively correlated to emotions ("joy" and "anger").

Table 6. Pearson correlation coefficient matrix among the sensory attributes, familiarity, liking and emotions for energy drinks.

Variables	Appearance	Aroma	Flavour	Sweetness	Bitterness	Mouthfeel	Aftertaste	Olike	Familiarity	Joy	Sadness	Disgust	Contempt	Anger	Fear	Surprise
Appearance	**1.00**	0.21	**0.39**	**0.33**	0.14	**0.32**	0.15	**0.36**	0.20	0.10	**−0.33**	0.14	0.11	0.18	0.01	0.20
Aroma	0.21	**1.00**	**0.56**	**0.32**	**0.38**	**0.34**	**0.39**	**0.53**	0.28	−0.14	0.02	−0.01	0.12	−0.11	0.03	−0.12
Flavour	**0.39**	**0.56**	**1.00**	**0.67**	**0.56**	**0.50**	**0.65**	**0.81**	**0.40**	−0.06	0.15	0.21	0.20	−0.11	−0.01	0.18
Sweetness	**0.33**	**0.32**	**0.67**	**1.00**	**0.38**	0.26	**0.59**	**0.57**	**0.38**	−0.19	−0.02	0.14	0.13	**−0.38**	0.03	0.02
Bitterness	0.14	**0.38**	**0.56**	**0.38**	**1.00**	**0.70**	**0.67**	**0.72**	**0.44**	−0.12	0.10	0.09	0.19	0.05	−0.03	0.08
Mouthfeel	**0.32**	**0.34**	**0.50**	0.26	**0.70**	**1.00**	**0.62**	**0.75**	**0.40**	−0.12	0.16	0.10	0.21	−0.11	−0.08	0.06
Aftertaste	0.15	**0.39**	**0.65**	**0.59**	**0.67**	**0.62**	**1.00**	**0.80**	**0.46**	−0.09	−0.07	0.11	0.15	**−0.29**	0.00	−0.03
Olike	**0.36**	**0.53**	**0.81**	**0.57**	**0.72**	**0.75**	**0.80**	**1.00**	**0.44**	−0.10	0.01	0.11	0.21	−0.10	0.06	0.09
Familiarity	0.20	0.28	**0.40**	**0.38**	**0.44**	**0.40**	**0.46**	**0.44**	**1.00**	−0.21	−0.07	−0.21	0.27	−0.14	0.13	0.18
Joy	0.10	−0.14	−0.06	−0.19	−0.12	−0.12	−0.09	−0.10	−0.21	**1.00**	−0.27	−0.02	−0.03	0.13	−0.07	0.18
Sadness	**−0.33**	0.02	0.15	−0.02	0.10	0.16	−0.07	0.01	−0.07	−0.27	**1.00**	0.13	−0.22	0.00	0.02	−0.08
Disgust	0.14	−0.01	0.21	0.14	0.09	0.10	0.11	0.11	−0.21	−0.02	0.13	**1.00**	−0.05	0.01	0.02	0.02
Contempt	0.11	0.12	0.20	0.13	0.19	0.21	0.15	0.21	0.27	−0.03	−0.22	−0.05	**1.00**	−0.08	−0.06	**0.80**
Anger	0.18	−0.11	−0.11	**−0.38**	0.05	−0.11	**−0.29**	−0.10	−0.14	0.13	0.00	0.01	−0.08	**1.00**	0.11	0.24
Fear	0.01	0.03	−0.01	0.03	−0.03	−0.08	0.00	0.06	0.13	−0.07	0.02	0.02	−0.06	0.11	**1.00**	0.04
Surprise	0.20	−0.12	0.18	0.02	0.08	0.06	−0.03	0.09	0.18	0.18	−0.08	0.02	**0.80**	0.24	0.04	**1.00**

Bold values show significant ($p < 0.05$) positive and negative correlations.

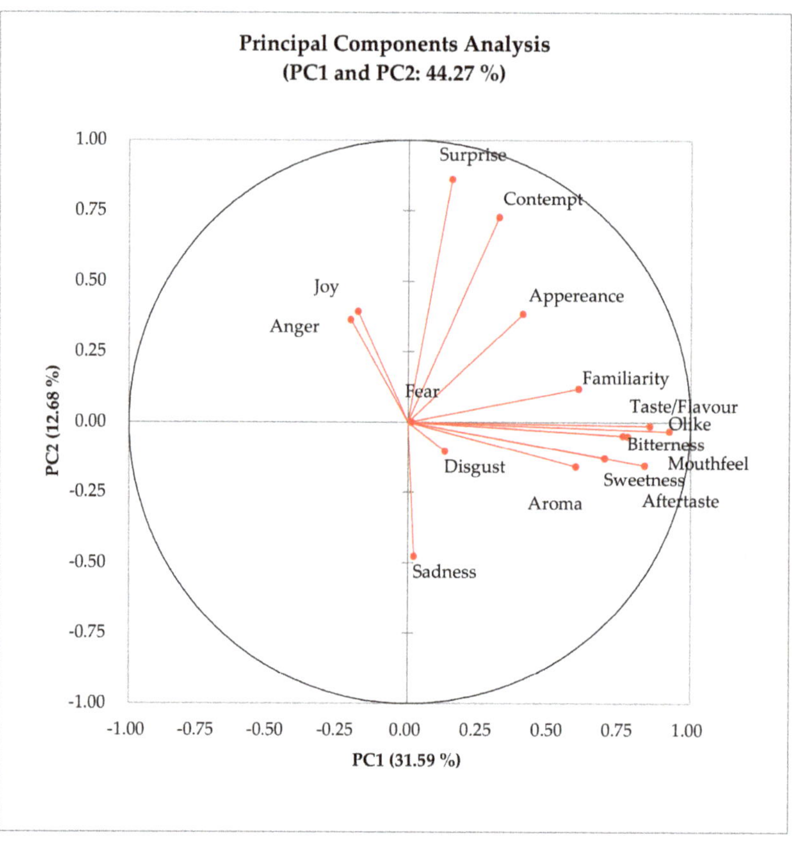

Figure 2. Principal components analysis (PCA) of the sensory attributes, familiarity, liking and emotions of energy drinks.

3.5. Comparison of Liking with Explicit and Implicit Emotions

The explicit and implicit emotions were examined using a linear regression model with overall liking as a response variable. The difference in regression coefficient and standard error for each explicit and implicit emotions were explained in Table 7. Model 1 shows the significant effect ($p < 0.05$) of explicit emotions on the overall liking of the energy drinks. The explicit emotions "good" and "satisfied" were positively related to overall liking with the regression coefficient values of 1.05 and 0.97, respectively. The explicit emotions "disgusted" and "surprised" were negatively related to overall liking with negative coefficients of -2.17 and -0.82, respectively. The relationship between the implicit emotions and overall liking were explained in Model 2. There was no significant effect ($p > 0.05$) of the emotions from the facial expressions on the overall liking of the samples. The implicit emotions "sadness" and "anger" were negatively associated with overall liking with the linear coefficient values of -23.67 and -43.63, respectively.

Table 7. A linear regression model * for each emotion method using overall liking as the response variable.

Model **	Emotion Parameter	Regression Coefficient	Standard Error (SE)	*p* Value	Regression Parameters		R^2
					Overall Mean	MSE	
Model 1 (self-reported emotions from CATA)	Active	0.51	0.34	0.14	5.75	1.87	0.59
	Adventurous	0.11	0.46	0.81			
	Bored	−0.15	0.63	0.81			
	Daring	0.87	0.74	0.24			
	Disgusted	*>−2.17*	*>0.58*	*><0.01*			
	Eager	−0.61	0.50	0.23			
	Energetic	−0.11	0.35	0.76			
	>Good	*>1.05*	*>0.32*	*><0.01*			
	Happy	0.59	0.40	0.15			
	Interested	0.39	0.40	0.34			
	Joyful	0.02	0.38	0.96			
	Mild	−0.36	0.41	0.38			
	Pleasant	0.27	0.35	0.44			
	>Satisfied	*>0.97*	*>0.35*	*>0.01*			
	Warm	0.15	0.50	0.76			
	Wild	−0.99	0.64	0.13			
	Anger	−0.14	1.21	0.91			
	>Surprised	*>−0.82*	*>0.40*	*>0.04*			
	Fear	−1.14	0.94	0.23			
	Contempt	−0.28	0.52	0.59			
Model 2 (facial expressions)	Joy	4.66	7.71	0.55	5.66	4.06	0.05
	Sadness	−23.67	52.73	0.66			
	Disgust	0.13	0.19	0.51			
	Contempt	8.95	10.72	0.41			
	Anger	−43.63	112.84	0.70			
	Fear	7.47	38.04	0.85			
	Surprised	−1.98	11.85	0.87			

* Data points were fitted using a linear regression model with overall liking as the response variable. MSE represents the mean square error value. R^2 is the coefficient of determination of the regression models. ** Bold italicized values indicate that the parameter was significantly different from 0 ($p < 0.05$).

4. Discussion

4.1. Traditional Method, Self-Reported Emotional Measurement and Purchase Intent

In general, energy drink A had higher liking scores than those of energy drink B. The high content of sugar and caffeine present in sample A might have influenced its overall liking. However, no significant differences were found between both samples in the liking of sensory attributes such as appearance, aroma, flavour, sweetness and aftertaste. The familiarity of the product or product category [54] and similar ingredients in both samples can be the reason for not having significant differences in the hedonic ratings of other attributes in the study. Earlier studies showed that the sensory profile of the products made with different ingredients had a strong influence on the hedonic liking scores [55] as compared to products made with similar ingredients.

For the self-elicited emotions, sample A had higher frequency values for positive emotions such as "active", "good", "adventurous", "pleasant", "joyful", "contempt", "warm", "interested", and "happy" compared to those of sample B. Sample A had a higher liking score, and also received a higher selection of high-arousal emotion terms such as "adventurous", and "active". This result shows that high-arousal emotional terms were important for brand liking in the energy drink category. The selection of high-arousal emotions such as "active" and "adventurous" can be due to the high sugar and caffeine contents in the energy drinks. Specterman et al. (2005) studied that the combined effect of caffeine and glucose had increased excitability and impulsiveness, as blood glucose level increased after consumption [56]. The explicit measurements showed that the positive emotions such as "happy", "joyful", and "pleasant" have higher frequency counts as compared to negative emotions such as "sad" and "angry". These results are in line with earlier findings that consumers use more positive emotions to describe food products than negative emotions [29,57]. In the present study, there was no significant difference in the purchase intent based on the sensory attributes of the energy drinks. Although, the majority of panellists (68%) preferred buying energy drink A rather than energy drink B (55%). This shows that explicit emotions and the overall liking of the product influence the purchase decisions taken by the consumers, as studied earlier [58,59] and plays an essential part in our lives.

4.2. Automated Facial Expression Response Measurement

No significant differences in the facial expressions were found. Small sample size could be a reason for the obtained results. Simultaneously, high individual variability in the identified emotions by AFER has previously been reported as an issue in discriminating products [60]. The quasi-absence of the emotions reported through AFER may be attributed to the liquid state of the test samples. Fewer facial movements involved with the liquid state [61,62], absence of apparent emotions [39] evoked by energy drinks or poor emotion recognition by the AFER and higher culture-to-culture (India, China, Cambodia, Vietnam, Korea and Hispanic respondents in this study) or individual-to-individual variances could be, to name a few, some of the reasons for marginal differentiation. Although neutral to positive emotions have been previously found to elicit a few facial expressions [37,63], the efficacy of AFER systems to differentiate between two competing products of a category was not sufficient to replace the existing traditional methods. Pragmatically, products competing in the same category, in general, are not profoundly different from each other, and a high negative valence associated with some of them is not expected. In such cases, the capacity of AFER to differentiate would be limited without the complement of traditional sensory techniques. Culturally specific display rules may also be a reason for the lack of differentiation. These rules govern the amplifying, dampening, or altogether masking of the facial expressions. The use of water, basic taste solutions at varying concentrations and target populations (culturally specific) for calibration or testing may be an option for the efficacy test of AFER systems. A large sample size study using both explicit and implicit methods may shed some light on the topic in the future, but practitioners should complement implicit methods with traditional methods for immediate applications. For future applications, implicit methods can have considerable implications in the retail and foodservice industry, and practitioners should keep testing this methodology.

In comparison to sample A, sample B evoked some negative facial expressions, such as nose scrunch, widen eyes, lip suck, which altogether represent disgust [64] and anger. c Contempt and disgust belong to the family of hostile emotions [65]. Limited facial movement hinders the precise measure of some of the facial expressions such as "fear", "sad" and "anger" [37,63]. Higher attention and engagement observed with sample A may imply that the sample was more enjoyable at the time of tasting. The caffeine, glucose and carbonation in energy drinks enhance the mood and the level of energetic arousal [43] and might be responsible for higher attention and engagement, irrespective of the sample type. More dimpler expressions, though not statistically significant, were observed for sample A. Dimpler has been previously identified as a predictor of positive emotion ratings by machine learning models [66]. In the case of energy drink A, the intensity of dimpler facial movements is higher as compared to energy drink B; thus, exhibiting the higher intensity of emotion contempt during the tasting of energy drink A. The participants elicited a higher intensity of joy and smile in sample B through implicit measures as compared to sample A, which was not the case using the nine-point scale ratings and the explicit study of emotions. Duchenne or genuine smiles are caused by the activation of facial action units 6 and 12 [67], and generally, this is the result of enjoyment and happiness. However, previous studies showed that a smile could be misleading, as many people smile as a sign of embarrassment [68], disappointment [69] or deliberately to hide emotions [70]. This shows that overall liking, explicit emotional responses, and implicit emotional responses vary in the outcome that they have in relationship to hedonic reactions. Explicit methods of emotion measure the conscious and cognitive actions or associations with the food product [14], whereas implicit methods measure the unconscious responses to the stimuli [71]. This finding was in accordance with the study, which stated that the overall liking, self-reported questionnaire and unconscious responses of the consumers are weakly to moderately correlated [37].

4.3. Multivariate Analysis of Hedonic and Facial Expression

Based on the Pearson correlation coefficient, familiarity influenced the overall liking of the product significantly. Similar findings were reported in earlier studies of meat [72,73], cheese [74], teas [75] and spirulina-filled pasta [76]. The effect of familiarity on consumers' preferences of the product also varies with socio-demographic factors [74] as well as cross-culturally [58]. Consumers feel more comfortable with the acquainted brands with which they are satisfied rather than exploring a new one. Familiarity and liking also affect the hedonic ratings of the sensory attributes (taste/flavour, aroma, texture, appearance, sweetness and bitterness) [75], as in the present study, the consumers gave higher liking ratings to the beverage product having familiar sensory attributes. In the case of implicit emotions, the emotion "anger" had been negatively correlated to sweetness and aftertaste. The participants felt less the negative emotion "anger" due to the sweetness and aftertaste of the product. Sweet foods elicited more positive emotions as compared to negative emotions in a previous study [77]. Overall, the results obtained from the self-reported and implicit reactions which provide meaningful insights to understand the differences in energy drinks. Thus, the correlation between emotions and sensory attributes can help food innovators to launch a promising product in a competitive market.

5. Conclusions

This preliminary study showed that the samples were significantly different based on overall liking. However, the sensory profiles of the energy drinks were not significantly different due to the similar product category, which also affected the purchase intent of the samples. In the case of explicit emotions, the magnitude of the self-reported positive emotions was slightly higher than negative emotions in both sample types. Higher self-reported positive emotions were reported for energy drink type A compared to B. However, in the case of implicit emotions by automated facial responses, positive emotions were expressed for both energy drinks (A and B). This study concludes that the traditional methods, self-reported emotional measurements and automated facial expression responses can vary in their outcome; however, all these reactions provide meaningful insights into the differentiation of the products. Future studies can be planned to test close- competitors in the same product category with explicit and implicit methods for efficacy-checking of the techniques based on the findings. Non-liquid food categories may produce better differentiation with implicit methods, but a study is needed to validate this assumption. Cultural manifestations, especially those of South and East Asia, in the case of facial expressions, are very subtle and offer another opportunity to test for. The test sensitivity of implicit methods may be more drastically affected with a small sample size; hence, power analysis for implicit methods should be revised. A connection between facial expressions and self-reported emotions could be another important area to study. Finally, based on the experimental findings in this study, it can be concluded that implicit emotion measurement methods are still in their preliminary stages and require many more investigations.

6. Limitations

A considerable proportion of the test population was of Asian origin, which would be a limitation in the generalization of these findings over other demographics. Additionally, the limited sample size may pose certain constraints in analysing and interpreting the results.

In addition, much recognition is required to understand the effect of the environment on emotions. Therefore, future studies are recommended to employ a greater number of participants and in different contexts such as central location tests and live experience or virtual reality sets to understand the environmental effect on the emotions depicted by facial expression.

Author Contributions: Conceptualization, D.D.T. and A.M.; methodology, D.D.T. and A.M.; formal analysis, A.M.; investigation, A.M., M.K., C.S., M.T., and D.D.T.; data curation, A.M., C.S., and D.D.T.; writing—original draft preparation, A.M.; writing—review and editing, A.M., D.D.T., C.S., and R.H.; supervision, D.D.T., C.S., and R.H.; project administration, R.H., and D.D.T.; funding acquisition, R.H., and D.D.T. All authors have read and agreed to the published version of the manuscript.

Funding: This research was funded and supported by Department of Wine, Food, and Molecular Biosciences, Faculty of Agriculture and Life Sciences, and the Centre of Excellence—Food for Future Consumers, Lincoln University, New Zealand.

Institutional Review Board Statement: The study was conducted according to the guidelines of the Declaration of Helsinki, and approved by the Human Ethics Committee of Lincoln University (Approval: 2019-68).

Informed Consent Statement: Informed consent was obtained from all subjects involved in the study.

Data Availability Statement: The data presented in this study are available on request from the corresponding author.

Acknowledgments: The authors would like to acknowledge all the subjects who participated in the sensory sessions.

Conflicts of Interest: The authors declare no conflict of interest. The funders had no role in the design of the study; in the collection, analyses, or interpretation of data; in the writing of the manuscript, or in the decision to publish the results.

References

1. Costa, A.I.; Jongen, W. New insights into consumer-led food product development. *Trends Food Sci. Technol.* **2006**, *17*, 457–465. [CrossRef]
2. Talavera, M.; Chambers, E., IV. Using sensory sciences help products succeed. *Br. Food J.* **2017**, *119*, 2130–2144. [CrossRef]
3. Torrico, D.D.; Hutchings, S.C.; Ha, M.; Bittner, E.P.; Fuentes, S.; Warner, R.D.; Dunshea, F.R. Novel techniques to understand consumer responses towards food products: A review with a focus on meat. *Meat Sci.* **2018**, *144*, 30–42. [CrossRef] [PubMed]
4. Civille, G.V.; Oftedal, K.N. Sensory evaluation techniques—Make "good for you" taste "good". *Physiol. Behav.* **2012**, *107*, 598–605. [CrossRef] [PubMed]
5. Claudia Gonzalez, V.; Damir, D.T.; Frank, R.D.; Sigfredo, F. Emerging Technologies Based on Artificial Intelligence to Assess the Quality and Consumer Preference of Beverages. *Beverages* **2019**, *5*, 62. [CrossRef]
6. Sharma, C.; Swaney-Stueve, M.; Severns, B.; Talavera, M. Using correspondence analysis to evaluate consumer terminology and understand the effects of smoking method and type of wood on the sensory perception of smoked meat. *J. Sens. Stud.* **2019**, *34*, e12535. [CrossRef]
7. Gonzalez Viejo, C.; Fuentes, S.; Torrico, D.; Howell, K.; Dunshea, F.R. Assessment of beer quality based on foamability and chemical composition using computer vision algorithms, near infrared spectroscopy and machine learning algorithms. *J. Sci. Food Agric.* **2018**, *98*, 618–627. [CrossRef]
8. Sharma, C.; Chambers, E., IV; Jayanty, S.S.; Sathuvalli Rajakalyan, V.; Holm, D.G.; Talavera, M. Development of a lexicon to describe the sensory characteristics of a wide variety of potato cultivars. *J. Sens. Stud.* **2020**, *35*, e12577. [CrossRef]
9. Kong, Y.; Sharma, C.; Kanala, M.M.; Thakur, M.; Li, L.; Xu, D.; Harrison, R.; Torrico, D. Virtual reality and immersive environments on sensory perception of chocolate products: A preliminary study. *Foods* **2020**, *9*, 515. [CrossRef] [PubMed]
10. Chetan, S.; Sastry, S.J.; Edgar, C.; Martin, T. Segmentation of Potato Consumers Based on Sensory and Attitudinal Aspects. *Foods* **2020**, *9*, 161. [CrossRef]
11. Damir, D.T.; Yitao, H.; Chetan, S.; Sigfredo, F.; Claudia Gonzalez, V.; Frank, R.D. Effects of Context and Virtual Reality Environments on the Wine Tasting Experience, Acceptability, and Emotional Responses of Consumers. *Foods* **2020**, *9*, 191. [CrossRef]
12. Peryam, D.R.; Pilgrim, F.J. Hedonic scale method of measuring food preferences. *Food Technol.* **1957**, *11*, 9–14.
13. Beckley, J.; Moskowitz, H.R.; Paredes, D. Beyond Hedonics: Looking at Purchase Intent and Emotions as Criterion Variables. 2008. Available online: http://ssrn.com/abstract=1439733 (accessed on 15 September 2020).

14. Köster, E.P.; Mojet, J. From mood to food and from food to mood: A psychological perspective on the measurement of food-related emotions in consumer research. *Food Res. Int.* **2015**, *76*, 180–191.

15. Cobb-Walgren, C.J.; Ruble, C.A.; Donthu, N. Brand Equity, Brand Preference, and Purchase Intent. *J. Advert.* **1995**, *24*, 25–40. [CrossRef]

16. McDuff, D.; El Kaliouby, R.; Cohn, J.F.; Picard, R.W. Predicting ad liking and purchase intent: Large-scale analysis of facial responses to ads. *IEEE Trans. Affect. Comput.* **2014**, *6*, 223–235. [CrossRef]

17. Patel, K.A.; Schlundt, D.G. Impact of moods and social context on eating behavior. *Appetite* **2001**, *36*, 111–118. [CrossRef]

18. Leigh Gibson, E. Emotional influences on food choice: Sensory, physiological and psychological pathways. *Physiol. Behav.* **2006**, *89*, 53–61. [CrossRef]

19. Berridge, K.; Winkielman, P. What is an unconscious emotion? (The case for unconscious "liking"). *Cogn. Emot.* **2003**, *17*, 181–211. [CrossRef]

20. Bettiga, D.; Lamberti, L.; Noci, G. Do mind and body agree? Unconscious versus conscious arousal in product attitude formation. *J. Bus. Res.* **2017**, *75*, 108–117. [CrossRef]

21. Fuentes, S.; Gonzalez Viejo, C.; Torrico, D.D.; Dunshea, F.R. Development of a Biosensory Computer Application to Assess Physiological and Emotional Responses from Sensory Panelists. *Sensors* **2018**, *18*, 2958. [CrossRef]

22. Gunaratne, T.M.; Gonzalez Viejo, C.; Fuentes, S.; Torrico, D.D.; Gunaratne, N.M.; Ashman, H.; Dunshea, F.R. Development of emotion lexicons to describe chocolate using the Check-All-That-Apply (CATA) methodology across Asian and Western groups. *Food Res. Int.* **2019**, *115*, 526–534. [CrossRef] [PubMed]

23. King, S.C.; Meiselman, H.L. Development of a method to measure consumer emotions associated with foods. *Food Qual. Prefer.* **2010**, *21*, 168–177. [CrossRef]

24. Gutjar, S.; De Graaf, C.; Kooijman, V.; de Wijk, R.A.; Nys, A.; Ter Horst, G.J.; Jager, G. The role of emotions in food choice and liking. *Food Res. Int.* **2015**, *76*, 216–223. [CrossRef]

25. Cardello, A.V.; Meiselman, H.L.; Schutz, H.G.; Craig, C.; Given, Z.; Lesher, L.L.; Eicher, S. Measuring emotional responses to foods and food names using questionnaires. *Food Qual. Prefer.* **2012**, *24*, 243–250. [CrossRef]

26. Chaya, C.; Pacoud, J.; Ng, M.; Fenton, A.; Hort, J. Measuring the Emotional Response to Beer and the Relative Impact of Sensory and Packaging Cues. *J. Am. Soc. Brew. Chem.* **2015**, *73*, 49–60. [CrossRef]

27. Mora, M.; Urdaneta, E.; Chaya, C. Emotional response to wine: Sensory properties, age and gender as drivers of consumers' preferences. *Food Qual. Prefer.* **2018**, *66*, 19–28. [CrossRef]

28. Ristic, R.; Danner, L.; Johnson, T.; Meiselman, H.; Hoek, A.; Jiranek, V.; Bastian, S. Wine-related aromas for different seasons and occasions: Hedonic and emotional responses of wine consumers from Australia, UK and USA. *Food Qual. Prefer.* **2019**, *71*, 250–260. [CrossRef]

29. Kanjanakorn, A.; Lee, J. Examining emotions and comparing the EsSense Profile® and the Coffee Drinking Experience in coffee drinkers in the natural environment. *Food Qual. Prefer.* **2017**, *56*, 69–79. [CrossRef]

30. Hu, X.; Lee, J. Emotions elicited while drinking coffee: A cross-cultural comparison between Korean and Chinese consumers. *Food Qual. Prefer.* **2019**, *76*, 160–168. [CrossRef]

31. Jack, R.E.; Caldara, R.; Schyns, P.G. Internal representations reveal cultural diversity in expectations of facial expressions of emotion. *J. Exp. Psychol. Gen.* **2012**, *141*, 19. [CrossRef]

32. Wendin, K.; Allesen-Holm, B.H.; Bredie, W.L. Do facial reactions add new dimensions to measuring sensory responses to basic tastes? *Food Qual. Prefer.* **2011**, *22*, 346–354. [CrossRef]

33. Hetherington, M.M. Understanding infant eating behaviour—Lessons learned from observation. *Physiol. Behav.* **2017**, *176*, 117–124. [CrossRef]

34. Oster, H. The repertoire of infant facial expressions: An ontogenetic perspective. In *Emotional Development: Recent Research Advances*; Nadel, J., Muir, D., Eds.; Oxford University Press: Oxford, UK, 2005; pp. 261–292.

35. Alaoui-Ismaili, O.; Vernet-Maury, E.; Dittmar, A.; Delhomme, G.; Chanel, J. Odor hedonics: Connection with emotional response estimated by autonomic parameters. *Chem. Senses* **1997**, *22*, 237–248. [CrossRef]

36. Zeinstra, G.G.; Koelen, M.; Colindres, D.; Kok, F.; De Graaf, C. Facial expressions in school-aged children are a good indicator of 'dislikes', but not of 'likes'. *Food Qual. Prefer.* **2009**, *20*, 620–624. [CrossRef]

37. Danner, L.; Haindl, S.; Joechl, M.; Duerrschmid, K. Facial expressions and autonomous nervous system responses elicited by tasting different juices. *Food Res. Int.* **2014**, *64*, 81–90. [CrossRef]

38. De Wijk, R.A.; Kooijman, V.; Verhoeven, R.H.; Holthuysen, N.T.; de Graaf, C. Autonomic nervous system responses on and facial expressions to the sight, smell, and taste of liked and disliked foods. *Food Qual. Prefer.* **2012**, *26*, 196–203. [CrossRef]

39. Mahieu, B.; Visalli, M.; Schlich, P.; Thomas, A. Eating chocolate, smelling perfume or watching video advertisement: Does it make any difference on emotional states measured at home using facial expressions? *Food Qual. Prefer.* **2019**, *77*, 102–108. [CrossRef]

40. Costa, B.M.; Hayley, A.; Miller, P. Young adolescents' perceptions, patterns, and contexts of energy drink use. A focus group study. *Appetite* **2014**, *80*, 183–189. [CrossRef]

41. Childs, E. Influence of energy drink ingredients on mood and cognitive performance. *Nutr. Rev.* **2014**, *72*, 48–59. [CrossRef]

42. Warburton, D.M. Effects of caffeine on cognition and mood without caffeine abstinence. *Psychopharmacology* **1995**, *119*, 66–70. [CrossRef]

43. Smit, H.; Cotton, J.; Hughes, S.; Rogers, P. Mood and cognitive performance effects of" energy" drink constituents: Caffeine, glucose and carbonation. *Nutr. Neurosci.* **2004**, *7*, 127–139. [CrossRef]

44. Peacock, A.; Martin, F.H.; Carr, A. Energy drink ingredients. Contribution of caffeine and taurine to performance outcomes. *Appetite* **2013**, *64*, 1–4. [CrossRef]

45. Warburton, D.M.; Bersellini, E.; Sweeney, E. An evaluation of a caffeinated taurine drink on mood, memory and information processing in healthy volunteers without caffeine abstinence. *Psychopharmacology* **2001**, *158*, 322–328. [CrossRef]

46. Birkás, B.; Dzhelyova, M.; Lábadi, B.; Bereczkei, T.; Perrett, D.I. Cross-cultural perception of trustworthiness: The effect of ethnicity features on evaluation of faces' observed trustworthiness across four samples. *Personal. Individ. Differ.* **2014**, *69*, 56–61. [CrossRef]

47. Ma, J.; Yang, B.; Luo, R.; Ding, X. Development of a facial-expression database of Chinese Han, Hui and Tibetan people. *Int. J. Psychol.* **2020**, *55*, 456–464. [CrossRef]

48. Jirangrat, W.; Wang, J.; Sriwattana, S.; No, H.K.; Prinyawiwatkul, W. The split plot with repeated randomised complete block design can reduce psychological biases in consumer acceptance testing. *Int. J. Food Sci. Technol.* **2014**, *49*, 1106–1111. [CrossRef]

49. Gacula, M., Jr.; Rutenbeck, S. Sample size in consumer test and descriptive analysis. *J. Sens. Stud.* **2006**, *21*, 129–145. [CrossRef]

50. Wild-Wall, N.; Dimigen, O.; Sommer, W. Interaction of facial expressions and familiarity: ERP evidence. *Biol. Psychol.* **2008**, *77*, 138–149. [CrossRef]

51. Danner, L.; Sidorkina, L.; Joechl, M.; Duerrschmid, K. Make a face! Implicit and explicit measurement of facial expressions elicited by orange juices using face reading technology. *Food Qual. Prefer.* **2014**, *32*, 167–172. [CrossRef]

52. Jones, D.V.; Work, C.E. Volume of a swallow. *Am. J. Dis. Child.* **1961**, *102*, 427. [CrossRef]

53. Lawless, H.T.; Bender, S.; Oman, C.; Pelletier, C. Gender, age, vessel size, cup vs. straw sipping, and sequence effects on sip volume. *Dysphagia* **2003**, *18*, 196–202.

54. Hersleth, M.; Ueland, Ø.; Allain, H.; Næs, T. Consumer acceptance of cheese, influence of different testing conditions. *Food Qual. Prefer.* **2005**, *16*, 103–110. [CrossRef]

55. Schouteten, J.J.; De Pelsmaeker, S.; Juvinal, J.; Lagast, S.; Dewettinck, K.; Gellynck, X. Influence of sensory attributes on consumers' emotions and hedonic liking of chocolate. *Br. Food J.* **2018**, *120*, 1489–1503. [CrossRef]

56. Specterman, M.; Bhuiya, A.; Kuppuswamy, A.; Strutton, P.; Catley, M.; Davey, N. The effect of an energy drink containing glucose and caffeine on human corticospinal excitability. *Physiol. Behav.* **2005**, *83*, 723–728. [CrossRef]

57. Desmet, P.M.A.; Schifferstein, H.N.J. Sources of positive and negative emotions in food experience. *Appetite* **2008**, *50*, 290–301. [CrossRef]

58. Torrico, D.D.; Fuentes, S.; Gonzalez Viejo, C.; Ashman, H.; Dunshea, F.R. Cross-cultural effects of food product familiarity on sensory acceptability and non-invasive physiological responses of consumers. *Food Res. Int.* **2019**, *115*, 439–450. [CrossRef]

59. Danner, L.; Ristic, R.; Johnson, T.E.; Meiselman, H.L.; Hoek, A.C.; Jeffery, D.W.; Bastian, S.E.P. Context and wine quality effects on consumers' mood, emotions, liking and willingness to pay for Australian Shiraz wines. *Food Res. Int.* **2016**, *89*, 254–265. [CrossRef]

60. Arnade, E.A. Measuring Consumer Emotional Response to Tastes and Foods through Facial Expression Analysis. Master's Thesis. Virginia Tech, Blacksburg, VA, USA, 2013.

61. Zhi, R.; Wan, J.; Zhang, D.; Li, W. Correlation between hedonic liking and facial expression measurement using dynamic affective response representation. *Food Res. Int.* **2018**, *108*, 237–245. [CrossRef]

62. Leitch, K.; Duncan, S.; O'keefe, S.; Rudd, R.; Gallagher, D. Characterizing consumer emotional response to sweeteners using an emotion terminology questionnaire and facial expression analysis. *Food Res. Int.* **2015**, *76*, 283–292. [CrossRef]

63. Vivi, T.; Yassierli, Y.; Hardianto, I. Basic Emotion Recogniton using Automatic Facial Expression Analysis Software. *J. Optimasi Sist. Ind.* **2019**, *18*, 55–64. [CrossRef]

64. Pochedly, J.T.; Widen, S.C.; Russell, J.A. What emotion does the "facial expression of disgust" express? *Emotion* **2012**, *12*, 1315. [CrossRef]

65. Miceli, M.; Castelfranchi, C. Contempt and disgust: The emotions of disrespect. *J. Theory Soc. Behav.* **2018**, *48*, 205–229. [CrossRef]

66. Haines, N. Decoding Facial Expressions that Produce Emotion Valence Ratings with Human-Like Accuracy. Master's Thesis. The Ohio State University, Columbus, OH, USA, 2017.

67. Ekman, P.; Friesen, W.; Hager, J. The Facial Action Coding Systems. In *Research Nexus eBook*, 2nd ed.; Weidenfeld & Nicolson: London, UK; Salt Lake City, UT, USA, 2002.

68. Weiss, J.; Ekman, P. *Telling Lies: Clues to Deceit in the Marketplace, Politics, and Marriage*; W. W. Norton & Company: New York, NY, USA, 2011.

69. Woodzicka, J.A.; LaFrance, M. Real versus imagined gender harassment. *J. Soc. Issues* **2001**, *57*, 15–30. [CrossRef]

70. Gunnery, S.D.; Hall, J.A.; Ruben, M.A. The deliberate Duchenne smile: Individual differences in expressive control. *J. Nonverbal Behav.* **2013**, *37*, 29–41. [CrossRef]

71. Mojet, J.; Dürrschmid, K.; Danner, L.; Jöchl, M.; Heiniö, R.-L.; Holthuysen, N.; Köster, E. Are implicit emotion measurements evoked by food unrelated to liking? *Food Res. Int.* **2015**, *76*, 224–232. [CrossRef]

72. Borgogno, M.; Favotto, S.; Corazzin, M.; Cardello, A.V.; Piasentier, E. The role of product familiarity and consumer involvement on liking and perceptions of fresh meat. *Food Qual. Prefer.* **2015**, *44*, 139–147. [CrossRef]

73. Prescott, J.; Young, O.; Zhang, S.; Cummings, T. Effects of added "flavour principles" on liking and familiarity of a sheepmeat product: A comparison of Singaporean and New Zealand consumers. *Food Qual. Prefer.* **2004**, *15*, 187–194. [CrossRef]
74. Nacef, M.; Lelièvre-Desmas, M.; Symoneaux, R.; Jombart, L.; Flahaut, C.; Chollet, S. Consumers' expectation and liking for cheese: Can familiarity effects resulting from regional differences be highlighted within a country? *Food Qual. Prefer.* **2019**, *72*, 188–197. [CrossRef]
75. Kim, Y.-K.; Jombart, L.; Valentin, D.; Kim, K.-O. Familiarity and liking playing a role on the perception of trained panelists: A cross-cultural study on teas. *Food Res. Int.* **2015**, *71*, 155–164. [CrossRef]
76. Grahl, S.; Strack, M.; Mensching, A.; Mörlein, D. Alternative protein sources in Western diets: Food product development and consumer acceptance of spirulina-filled pasta. *Food Qual. Prefer.* **2020**, *84*. [CrossRef]
77. Kim, J.-Y.; Prescott, J.; Kim, K.-O. Emotional responses to sweet foods according to sweet liker status. *Food Qual. Prefer.* **2017**, *59*, 1–7. [CrossRef]

Article

Beer and Consumer Response Using Biometrics: Associations Assessment of Beer Compounds and Elicited Emotions

Claudia Gonzalez Viejo [1,*,†], Raúl Villarreal-Lara [2,3,†], Damir D. Torrico [4],
Yaressi G. Rodríguez-Velazco [2], Zamantha Escobedo-Avellaneda [2], Perla A. Ramos-Parra [2],
Ronit Mandal [5], Anubhav Pratap Singh [5], Carmen Hernández-Brenes [2] and
Sigfredo Fuentes [1]

[1]   Digital Agriculture, Food and Wine Sciences Group, School of Agriculture and Food, Faculty of Veterinary
      and Agricultural Sciences, University of Melbourne, Melbourne, VIC 3010, Australia;
      sfuentes@unimelb.edu.au
[2]   Tecnologico de Monterrey, Escuela de Ingeniería y Ciencias, Ave. Eugenio Garza Sada 2501,
      Monterrey 64849, N.L., Mexico; raulvl@tec.mx (R.V.-L.); A00801965@itesm.mx (Y.G.R.-V.);
      zamantha.avellaneda@tec.mx (Z.E.-A.); perlaramos@tec.mx (P.A.R.-P.); chbrenes@itesm.mx (C.H.-B.)
[3]   SensoLab Solutions, Centro de Innovación y Transferencia Tecnológica (CIT2), Ave. Eugenio Garza Sada
      #427 Col. Altavista, Monterrey 64849, N.L., Mexico
[4]   Department of Wine, Food and Molecular Biosciences, Faculty of Agriculture and Life Sciences,
      Lincoln University, Lincoln 7647, New Zealand; damir.torrico@lincoln.ac.nz
[5]   Food, Nutrition, and Health, Faculty of Land and Food Systems, University of British Columbia, 2205,
      East Mall, Vancouver, BC V6T 1W4, Canada; ronit123@mail.ubc.ca (R.M.); anubhav.singh@ubc.ca (A.P.S.)
*   Correspondence: cgonzalez2@unimelb.edu.au
†   These authors contributed equally.

Received: 4 June 2020; Accepted: 17 June 2020; Published: 22 June 2020

Abstract: Some chemical compounds, especially alcohol, sugars, and alkaloids such as hordenine, have been reported as elicitors of different emotional responses. This preliminary study was based on six commercial beers selected according to their fermentation type, with two beers of each type (spontaneous, bottom, and top). Chemometry and sensory analysis were performed for all samples to determine relationships and patterns between chemical composition and emotional responses from consumers. The results showed that sweeter samples were associated with higher perceived liking by consumers and positive emotions, which corresponded to spontaneous fermentation beers. There was high correlation ($R = 0.91$; $R^2 = 0.83$) between hordenine and alcohol content. Beers presenting higher concentrations of both, and higher bitterness, were related to negative emotions. Further studies should be conducted, giving more time for emotional response analysis between beer samples, and comparing alcoholic and non-alcoholic beers with similar styles, to separate the effects of alcohol and hordenine. This preliminary study was a first attempt to associate beer compounds with the emotional responses of consumers using non-invasive biometrics.

Keywords: hordenine; happiness; beer consumption; sensory analysis; beer styles

---

1. Introduction

Beer is a complex alcoholic beverage in terms of its chemical composition and ingredients, such as barley, yeast, hops, and, in some beer products, includes adjuncts that may consist of other cereals or fruits [1,2]. The wide range of combinations that may be used from each of the ingredients, along with the differences in brewing methods, have a great influence on the development of beer's chemical and aroma profiles. Among the most important beer quality compounds are the iso-alpha acids from hops, which are responsible for the bitterness characteristic of the final product, proteins from yeast and barley, which contribute to the foamability and foam stability, and alcohol/sugar content, which determine the strength of beer [3–5]. Furthermore, beer also contains inorganic salts, alkaloids, polyphenols, aminoacids, and hop resins, which affect the physical and sensory characteristics of the final product [6]. Beer is a beverage product that potentially affects human emotional responses elicited by its chemical components.

Some chemical compounds in foods have been associated with either the suppression or release of certain neurotransmitters that trigger different emotions in humans. Most of these components are alkaloids, which are mainly considered as biological amines, but may present a diversity of structures in the form of esters or amides, or combined with sugars [7,8]. Cacao contains alkaloids such as theobromine, caffeine, phenylethylamine, and salsolinol, which have been studied for their psychoactive effects on humans. The synergistic effect of theobromine and caffeine has been associated with changes in mood such as energetic arousal and an increase in cognitive function, as well as changes in physiological responses such as heart rate [9,10]. Phenylethylamine is a compound similar to amphetamines, which triggers serotonin that regulates the mood and has been associated with emotions such as joy, happiness, and love [11,12]. Furthermore, salsolinol is able to bind with dopamine receptors and stimulate the release of endorphins, which leads to a sensation of reward and suppression of pain [9]. On the other hand, other chemical compounds, such as alcohol, cause a reduction in serotonin levels, which leads to depression [13].

Beer also contains some alkaloids, such as salsolinol and hordenine, the latter being in higher concentrations than salsolinol [14,15]. Hordenine is naturally found in barley during its germination; therefore, it is passed through the beer process in malting [16]. It has been reported that the hordenine concentration in beer mainly depends on the time, temperature, and humidity during the germination process of barley [17]. Even though it is found in low to moderate concentrations in beers, this alkaloid contributes to the diuretic effect of beer and provides some bitterness characteristics [16,18]. Regarding the effects of hordenine in humans, it has been reported to increase heart rate and blood pressure [19] as well as to stimulate the release of dopamine, which has been related to happiness [15,20]. On the other hand, other authors have concluded that beer flavors are the main factor responsible for dopamine release [21]. However, first study [15] was conducted by testing hordenine in radioligand assays, but not testing the compound in beer, while the second study [21] consisted in performing a positron emission tomography (PET) when tasting beer flavors. Hence, those experiments did not evaluate the effects of beer on consumers' emotional responses. Thus, there is still a large gap in the understanding of the mechanisms behind beer tasting and consumer perceptions.

This study aimed to assess the effect of beer compounds on the emotional responses of consumers using traditional sensory tests (self-reported responses) as well as non-invasive biometrics (unconscious responses). For these purposes, six beer samples from different fermentation types were used to measure the physicochemical data such as color, iso-alpha acids, hordenine, alcohol content, and bitterness, among others. Furthermore, a sensory session was conducted with Mexican beer consumers to obtain both self-reported and subconscious responses in order to assess their acceptability and elicited emotions. Multivariate data analysis was conducted to assess the relationship between the physicochemical, liking, and emotional responses from consumers.

2. Materials and Methods

Six commercial beer samples, two from each of the three different types of fermentation (Table 1), were selected from a pool of 24 beers previously analyzed for physicochemical and sensory descriptors [1,3,22–25]. The number of samples was limited to six as this is the recommended maximum number to avoid consumers' fatigue, especially due to alcohol content and bitterness, and this may also lead to a decrease in the quality of the responses [26,27]. Samples were purchased from a local supplier (Beer for Us S.A. de C.V., Monterrey, NL, México) and stored as described. For physicochemical characterization, samples were divided into aliquots and stored at −80 °C until their use, while samples for sensory evaluations were kept first at room temperature and then put in refrigeration (4 °C) 24 h before tests were conducted. Samples were defrosted at 4 °C and, prior to physicochemical analysis, were degassed using an ultrasound bath (SH30H, Elmasonic, Frechen, Germany) at 37 kHz for 30 min.

Table 1. Beer styles and labels of samples used to report results.

Beer Style	Beer Fermentation	Country of Origin	Label
Lambic Kriek	Spontaneous	Belgium	LK
Lambic Framboise	Spontaneous	Belgium	LF
Pale Lager	Bottom	Mexico	C
Pale Lager	Bottom	Mexico	H
Blonde Ale	Top	Belgium	L
Porter	Top	Poland	Z

2.1. Physicochemical Characterization

The color parameters (L*, a*, b*) were measured using a LabScan XE System (Hunter Associates Laboratory, Inc., Reston, VA, USA) colorimeter by triplicate using a CIELAB system. Instrumental color was expressed as the Hue angle (Equation (1)), Chroma (Equation (2)), and Yellowness Index (YI, Equation (3); [28]). Evaluation of viscosity was determined in triplicates using a rheometer (MCR 302, Anton Paar Canada Inc, Quebec, Canada) at 25 °C. All samples (30–40 mL) were measured for pH at 25 °C in triplicates using a Fisherbrand Accumet® AB15 pH meter (Fischer Scientific, Hampton, NH, USA) calibrated against standard buffers (3-point calibration: pH 4.00–0.1 M Potassium hydrogen phthalate buffer, pH 7.00—Potassium Phosphate Monobasic/Sodium Hydroxide buffer, pH 10.00–Potassium Carbonate/Potassium Tetraborate/Potassium Hydroxide/Disodium EDTA Dihydrate buffer). Titratable acidity (TA, as % acetic acid) was measured in duplicates, with the method described by Okafor et al. [29], and the following modifications. A total of 5 mL was titrated with 0.1 M NaOH until pH = 7.0 was achieved using bromothymol blue as an indicator.

$$\text{Hue angle} = \arctan\left(\frac{b^*}{a^*}\right) \times \frac{180}{3.14} \tag{1}$$

$$\text{Chroma} = \left(a^{*2} + b^{*2}\right)^{0.5} \tag{2}$$

$$YI = 142.86 \times \frac{b*}{L*} \tag{3}$$

The density of samples was assessed based on weight and volume (50 mL). Total dissolved solids (TDS) were measured in triplicates using a Yuelong YL-TDS2-A digital water quality tester (Zhengzhou Yuelong Electronic Technology Co., Ltd., Zhengzhou City, Henan Province, China). Salt concentration was obtained using two drops of the sample in triplicates added to a digital salt-meter (PAL-SALT Mohr, Atago Co., Ltd. Saitama, Japan). On the other hand, alcohol content was assessed using 18 mL of the sample at room temperature (20 °C) injected to an Alcolyzer Wine M alcohol meter (Anton Paar GmbH, GRAZ, Austria) with the wine extension method found in the equipment settings; the instrument has a maximum error of 0.1% vv^{-1}.

2.2. Characterization of Simple Sugars by HPLC-Refractive Index

The simple sugars profile was measured as described by Heredia-Olea et al. [30] and Alonso-Gómez et al. [31] with slight modifications. The samples were filtered through a polyvinylidene fluoride (PVDF) syringe filter (0.2 μm) and injected into high-performance liquid chromatography (HPLC) equipment (Waters HPLC Breeze model, Waters, Milford, MA, USA) with a refractive index detector

(Waters 2414) kept at 50 °C. The chromatographic separation was achieved using an ion-exclusion column Phenomenex Rezex ROA-organic acid h+ (250 × 4.6 mm, 8 µm particle size, Phenomenex, Torrance, CA, USA) at 60 °C. The mobile phase consisted of a 5 mM H_2SO_4 solution with a 20 min isocratic flow rate of 0.4 mL min^{-1} and with an injection volume of 10 µL. Glucose, maltose, and fructose quantifications were performed with calibration curves of HPLC-grade standards (Sigma-Aldrich, St. Louis, MO, USA).

Calibration curves for glucose (1–25 mg mL^{-1}), maltose (0.5–5 mg mL^{-1}), and fructose (1–25 mg mL^{-1}) of HPLC-grade standards (Sigma-Aldrich, St. Louis, MO, USA) were constructed for quantification purposes (data not shown).

2.3. Determination of Bitterness

Bitterness was assessed by manual isooctane extraction as described in the American Society of Brewing Chemists (ASBC) Methods of Analysis with the following modifications [32]. A total of 5 mL of beer was acidified with hydrochloric acid (HCl; 0.5 mL, 3M) and isooctane (10 mL); subsequently, it was homogenized for 15 min using a mechanical shaker. The separation of organic and aqueous layers was performed by centrifugation at 400 g × 5 min. Finally, the isooctane phase (upper) was measured spectrophotometrically at 275 nm. A calculation of bitterness units (IBU) of beer was obtained, as shown in Equation (4).

$$IBU = absorbance_{275} \times 50 \tag{4}$$

2.4. Characterization of Iso-α-Acids

The iso-α-acids profile was analyzed as described by Vanhoenacker et al. [33]. The beer samples (~25 mL) were filtered through a polytetrafluoroethylene (PTFE) syringe filter (0.2 µm) and directly injected to an Acquity Ultra-Performance Liquid Chromatography (UPLC, Waters, Milford, MA, USA) coupled to Diode-Array Detector (DAD), monitoring at 270 nm. The chromatographic separation was achieved using a Zorbax Extend C-18 column (100 × 3 mm, 3.5 µm particle size, Agilent, Santa Clara, SA, USA) kept at 35 °C. The mobile phase consisted of 5 mM ammonium acetate in 20% ethanol (pH 9.95) as phase A; acetonitrile/ethanol (60:40 v/v) as phase B. The solvent flow rate was 0.4 mL min^{-1}, using gradients of 0–3 min, 0% B; 3–4 min, 0–16% B; 4–54 min, 16–40% B; 54–57 min, 40–95% B; 57–65 min, 95%B; 65–67 min, 95–0% B, followed by 20 min re-equilibration. The iso-α-acids quantification was performed with calibration curves of commercial standards of trans-iso-α-acids in dicyclohexylamine (DCHA) obtained from the American Society of Brewing Chemists (Saint Paul, MN, USA). The standards were prepared in methanol (0.05% H_3PO_4) according to the supplier's recommendations and the ASBC Methods of Analysis [34].

2.5. Hordenine Determination by UPLC-MS/MS

Hordenine sample preparation was performed as described by Sommer et al. [35] with slight modifications. Beer samples were centrifuged for 15 min at 12,000× g and 4 °C; two dilution steps were followed. *Dilution I (Dil. I):* 50 µL of degassed beer were added to 450 µL of 0.1% formic acid. *Dilution II (Dil. II):* 20 µL of *Dil. I* were added to 980 µL of 0.1% formic acid. The solutions obtained after *Dil. II* were passed through a PVDF filter (0.2 µm, Thermo Scientific™, Waltham, MA, USA) prior to the analysis. For quantification, a calibration curve with a range of 0–0.1 ppm was developed using a stock solution (2 mg mL^{-1}) of a hordenine commercial standard (Sigma-Aldrich, St. Louis, MO, USA) prepared in formic acid (0.1%).

Hordenine separation and quantification were conducted in a Quattro Premier XE Micromass UPLC-MS/MS system (Waters, Milford, MA, USA) equipped with a triple quadrupole mass spectrometer (QQQ-MS) connected to an Acquity UPLC (Waters, Milford, MA, USA) with electrospray ionization (ESI) source in positive mode. Hordenine was analyzed in the multiple reaction monitoring (MRM) mode of m/z 165.95:121. Masslynx 4.1 software (Waters, Milford, MA, USA.) was used for data acquisition and instrument control. Hordenine separation was performed using a high strength silica (HSS T3 C18) column (2.1 mm × 100 mm, 1.8 µm particle size) coupled with a VanGuard HSS T3 C18 column (2.1 × 5 mm, 1.8 µm) maintained at 50 °C. Mobile phases were 0.1% formic acid in water (solvent A) and 0.1% formic acid in 70% acetonitrile and 30% methanol (solvent B) with 6.6 min total gradient solution as follows: 0–1 min, 5–15% B; 1–2 min, 15–40% B; 2–3 min, 40–70% B; 3–3.5 min, 70–100% B; 3.5–5 min, 100% B; 5–5.2 min, 100–5% B, followed by 1.4 min re-equilibration. The flow rate was kept constant at 0.5 mL min^{-1} with an injection volume of 10 µL. Nitrogen was used as the desolvation gas (400 L/h). The selected ion monitoring conditions were set as capillary voltage 2.5 kV, source temperature 120 °C, and desolvation temperature 400 °C. All determinations were conducted in triplicates.

2.6. Consumer Sensory Evaluation and Biometrics

A sensory session was carried out in Monterrey, NL, Mexico, which is the state with the highest alcoholic drinks consumption with beer as the leader [36,37]. The session was conducted with $N = 61$ beer consumers (frequency > three times a month; 54% males; 46% females) between 18 and 51 years old (mean age 25.6 ± 6.9 years). Participants were recruited via email and asked to participate in a graduate research project from the Department of Bioengineering, School on Engineering and Sciences of Tecnológico de Monterrey, Campus Monterrey, Mexico (Ethics ID: CSERDBT-0002). According to the Power analysis conducted using the Power and Sample Size Calculator from the SigmaXL ver. 8.15 software (SigmaXL Inc., Kitchener, ON, Canada), the number of participants was sufficient to find significant differences (1-β = 0.98) among the beer samples. The session was conducted at SensoLab Solutions SC, a sensory and consumer science laboratory center, located at the Technology Transfer and Innovation Center of Tecnológico de Monterrey, Mexico. The laboratory was equipped with eight individual sensory booths with uniform lighting. Each booth had an Android® (Google, Mountain View, CA, USA) Samsung Galaxy Tab 4 tablet (Samsung, Seoul, South Korea) displaying the Bio-Sensory application (App; The University of Melbourne, Parkville, Vic, Australia). The App was able to present the questionnaire (Table 2) and record videos from the participants while tasting the beer samples to further analyze their emotional responses [29]. Samples (30 mL) were served at refrigeration temperature (4 °C), and water was used as palate cleanser before and between each sample. To assess the visual descriptors of the beers, a video showing the pouring of the sample using the RoboBEER (The University of Melbourne, Parkville, Vic, Australia) was displayed in the App to avoid bias from the variability due to the pouring method and glass effects [22]. As shown in Table 2, two overall liking ratings were obtained at the start and end of the tasting to verify if there is a bias on this descriptor based on the evaluation of specific attributes.

Table 2. Questionnaire presented in the Bio-Sensory application.

Question/Descriptor	Answers (Options)	Scale
Overall liking (rated at the start of the test)	Dislike extremely—Like extremely	15-cm non-structured scale
Foam stability	Dislike extremely—Like extremely	15-cm non-structured scale
Foam height	Dislike extremely—Like extremely	15-cm non-structured scale
Bitterness	Dislike extremely—Like extremely	15-cm non-structured scale
Sweetness	Dislike extremely—Like extremely	15-cm non-structured scale
Acidity	Dislike extremely—Like extremely	15-cm non-structured scale
Aroma	Dislike extremely—Like extremely	15-cm non-structured scale
How do you feel when tasting this sample?		Face Scale (0–100)

Table 2. *Cont.*

Question/Descriptor	Answers (Options)	Scale
Check all emojis that depict how you feel when tasting this sample		Check all that apply (CATA)
Check all emotions that depict how you feel when tasting this sample	Active/Joyful/Aggressive/Bored/Affectionate/ Disgusted/Free/Friendly/Happy/Adventurous/ Guilty/Nostalgic/Calm/Pleasant/Satisfied/ Secure/Surprised/Worried *	Check all that apply (CATA)
Overall liking (rated at the end of the test)	Dislike extremely—Like extremely	15-cm non-structured scale

* Emotion-terms obtained from EsSense Profile® [38].

Videos were analyzed using an application developed based on the Affectiva software development kit (SDK; Affectiva, Boston, MA, USA). This application uses the histogram of the oriented gradient to detect and track the micro- and macro-movements of face features and is able to evaluate all videos in batch. Furthermore, it is capable of assessing facial expressions using support vector machine algorithms to translate them into emotions such as (i) contempt, (ii) disgust, (iii) sadness, (iv) surprise, (v) joy, (vi) valence, (vii) engagement, and (viii) attention, as well as emojis related to facial expressions such as (ix) smiley , (x) relaxed , (xi) winking face , (xii) stuck out tongue , (xiii) flushed , (xiv) rage , (xv) smirk , and (xvi) disappointed [39].

2.7. Statistical Analysis

All data were analyzed through ANOVA and least significant differences (LSD) as a post-hoc test ($\alpha = 0.05$) using Minitab 17.2.1 (Minitab Inc., State College, PA, USA). A linear correlation analysis was conducted for alcohol and hordenine values using Microsoft Excel (Microsoft, Redmond, WA, USA). Chemical, sensory (self-reported), and biometric responses were assessed using multivariate data analysis based on principal components analysis (PCA), and multiple factor analysis (MFA) with a customized code written in Matlab® R2019b (Mathworks, Inc., Natick, MA, USA) and XLSTAT ver. 2020.1.1 (Addinsoft Inc., New York, NY, USA), respectively.

3. Results

3.1. Physicochemical Results

Table 3 shows the mean values and results from the ANOVA for selected physicochemical parameters. There were significant differences ($p < 0.05$) between samples for all parameters. Sample Z had the lowest mean value for L* (26.58) as this is the darkest beer, while C had the highest value (59.36). Similarly, the yellow index (YI) was higher for Z (200.40) than all other samples, C being the lowest (15.69). Spontaneous fermentation beers (LK and LF) were the highest in density (1.02 and 1.03 g mL^{-1}, respectively), and significantly different from the other samples. On the other hand, LK was the most viscous (2.16 mPa s), followed by Z and H (1.80 mPa s), with C as the least viscous (1.48 mPa s). On the other hand, the spontaneous fermentation samples were the most acidic (LF: pH = 2.94, TA = 0.32; LK: pH = 3.17, TA = 0.41), while Z was the least acidic (pH = 4.42, TA = 0.17).

Figure 1 shows the means and ANOVA results of the total sugars, bitterness (Figure 1a), iso-alpha acids, and hordenine (Figure 1b). The spontaneous fermentation beers had significantly higher ($p < 0.05$) total sugar content (LF: 31.23 mg mL^{-1}; LK: 27.53 mg mL^{-1}) than the samples from other types of fermentation; for C, the sugar concentration was non-detectable with the chromatographic conditions used. Sample Z was the highest in both bitterness (34.98 IBU) and total iso alpha-acids (21.41 mg L^{-1}), while LF was the least bitter (bitterness: 5.08 IBU; total iso-alpha acids: 0.60 mg L^{-1}). On the other hand, the top fermentation beers (Z and L) had the highest concentrations of hordenine

(Z: 4.24 mg L^{-1}; L: 3.22 mg L^{-1}), while spontaneous fermentation sample LF had the lowest content (0.98 mg L^{-1}).

Table 3. Physicochemical characterization of commercial beers.

Sample	Color						Density (g mL^{-1})	Viscosity (mPa s)	pH	Titratable Acidity
	L*	a*	b*	Hue	Chroma	YI				
LK	36.70 d† ± 0.12	23.68 b ± 0.08	17.76 c ± 0.05	0.64 b ± 0.001	29.60 c ± 0.09	69.15 c ± 0.39	1.02 a ± 0.001	2.16 a ± 0.11	3.17 d ± 0.01	0.41 a ± 0.01
LF	29.67 e ± 0.08	26.54 a ± 0.16	20.16 b ± 0.32	0.65 b ± 0.005	33.33 b ± 0.32	97.11 b ± 1.79	1.03 a ± 0.002	1.73 bc ± 0.06	2.94 e ± 0.01	0.32 b ± 0.03
C	59.36 a ± 0.07	−1.27 d ± 0.01	6.52 f ± 0.04	−1.38 c ± 0.003	6.64 f ± 0.03	15.69 f ± 0.07	1.00 c ± 0.003	1.48 d ± 0.09	4.29 b ± 0.00	0.11 d ± 0.00
H	58.72 b ± 0.25	−1.21 d ± 0.03	8.99 e ± 0.09	−1.44 d ± 0.002	9.07 e ± 0.09	21.87 e ± 0.21	1.00 c ± 0.002	1.80 b ± 0.07	4.31 b ± 0.01	0.10 d ± 0.01
L	56.68 c ± 0.27	−1.02 d ± 0.01	16.25 d ± 0.08	−1.51 e ± 0.001	16.28 d ± 0.08	40.96 d ± 0.09	1.01 b ± 0.002	1.54 cd ± 0.02	4.24 c ± 0.01	0.11 d ± 0.00
Z	26.58 f ± 0.16	16.82 c ± 0.14	37.28 a ± 0.52	1.14 a ± 0.003	40.90 a ± 0.53	200.40 a ± 1.59	1.00 c ±0.003	1.80 b ± 0.00	4.42 a ± 0.01	0.17 c ± 0.01

Abbreviations: CIELAB color parameters (L*: lightness, a*: red/green, b*: blue/yellow), YI: yellowness index. † Values represent the mean ± standard error ($n_{Titratable\ Acidity} = 2$, $n_{Color,\ Density,\ Viscosity,\ pH} = 3$). Abbreviations of samples may be found in Table 1. Different letters within a column indicate that values are significantly different according to the least significant difference test (LSD; $p < 0.05$).

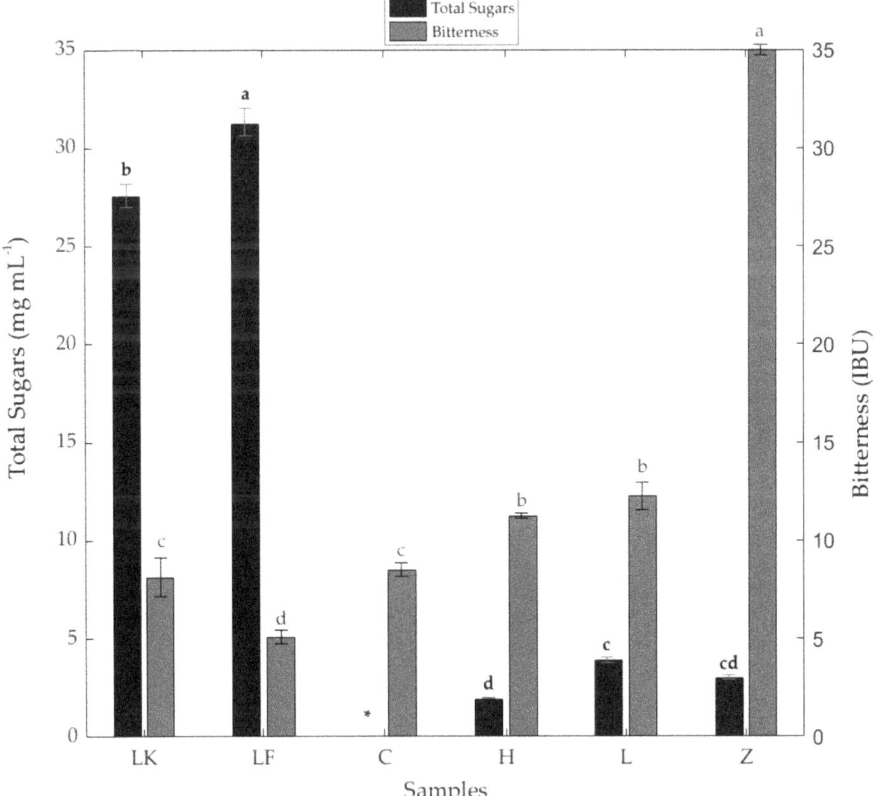

(**a**)

Figure 1. *Cont.*

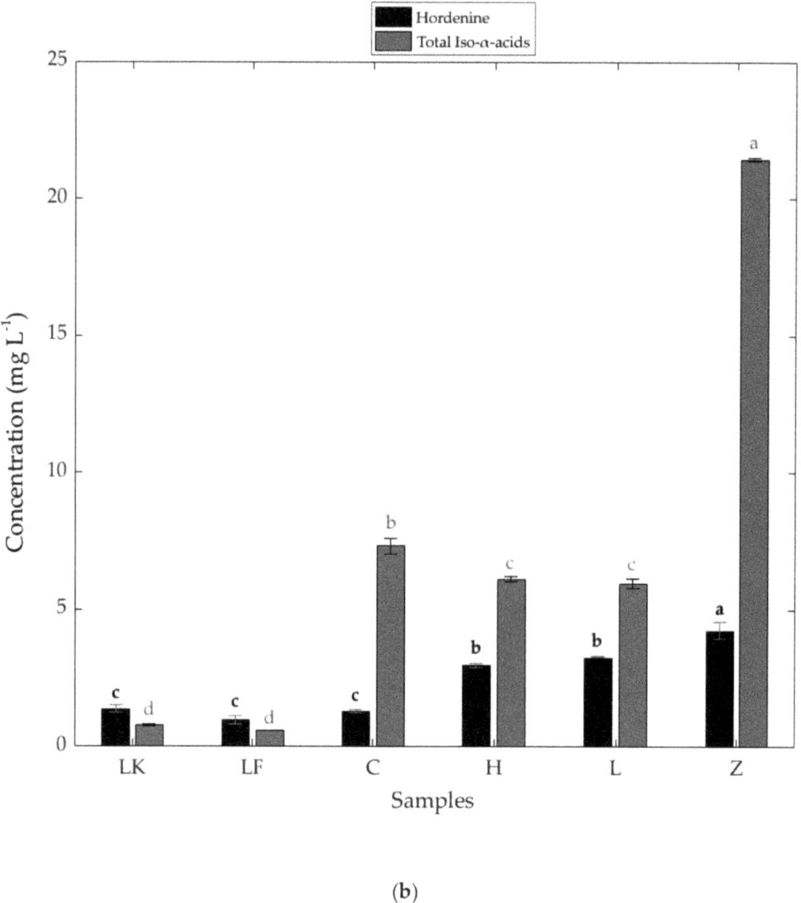

(**b**)

Figure 1. Chemical characterization of commercial beers, including (**a**) total sugars (mg mL^{-1}), bitterness (IBU), (**b**) hordenine (mg L^{-1}), total Iso-α-acid concentration (mg L^{-1}). Different letters above bars denote significant differences between beer samples, for the same chemical parameter, according to the least significant difference test (LSD; $p < 0.05$). * Total sugars not detected in beer C. All values are the mean ± SE (error bars) of independent determinations. $n = 3$, hordenine, and bitterness; $n = 2$, total sugars, and total-α-acids. Abbreviations of samples may be found in Table 1.

Table 4 shows that the simple sugars from the spontaneous fermentation samples (LF and LK) were mainly composed of glucose (LF: 14.32 mg mL^{-1}; LK: 13.91 mg mL^{-1}), followed by fructose (LF: 13.51 mg mL^{-1}; LK: 12.56 mg mL^{-1}), and maltose (LF: 3.40 mg mL^{-1}; LK: 1.06 mg mL^{-1}). Sample H had higher values of maltose (0.79 mg mL^{-1}) than glucose (0.60 mg mL^{-1}) and fructose (0.50 mg mL^{-1}), while L was higher in fructose (2.04 mg mL^{-1}) than glucose (1.87 mg mL^{-1}) and did not contain maltose. Spontaneous fermentation beers were the highest in salt concentration (LK and LF: 0.10%), while C was the lowest (0.05%). A similar trend was found for TDS with LF and LK; although being significantly different, both presented the highest values (LF: 1226 ppm; LK: 1148 ppm), while C had the lowest with 658 ppm. Top fermentation beers showed the highest alcohol content (Z: 9.47%; L: 6.68%), while spontaneous fermentation samples had the lowest (LF: 2.53%; LK: 3.53%). A similar trend was found for the content of trans-Isocohumulone and trans-Isohumulone parameters with Z being the highest concentration (10.95 mg L^{-1}, and 10.46 mg L^{-1}, respectively), and LF the lowest (0.22 mg L^{-1}, and 0.38 mg L^{-1}, respectively).

Table 4. Simple sugars, salt, total dissolved solids, ethanol content, and iso-α-acids of commercial beers.

Sample	Simple Sugars (mg mL^{-1})			Salt (%)	Total Dissolved Solids (ppm)	Alcohol Content (%)	Iso-α-Acids (mg L^{-1})	
	Glucose	Fructose	Maltose				Trans-Isocohumulone	Trans-Isohumulone
LK	13.91 a* ± 0.24	12.56 b ± 0.31	1.06 c ± 0.04	0.10 a ± 0.00	1148.00 b ± 11.00	3.53 e ± <0.001	0.33 e ± 0.01	0.45 d ± 0.01
LF	14.32 a ± 0.62	13.51 a ± 0.01	3.40 a ± 0.07	0.10 a ± 0.00	1226.00 a ± 7.00	2.53 f ± <0.001	0.22 e ± 0.01	0.38 d ± 0.01
C	ND	ND	ND	0.05 e ± 0.00	658.00 f ± 9.61	4.62 d ± <0.001	3.44 b ± 0.08	3.91 b ± 0.22
H	0.60 c ± 0.00	0.50 d ± 0.00	0.79 d ± 0.03	0.06 d ± 0.00	738.00 e ± 4.04	4.97 c ± <0.001	2.81 c ± 0.00	3.27 c ± 0.11
L	1.87 b ± 0.06	2.04 c ± 0.08	0.00 e ± 0.00	0.07 c ± 0.00	898.67 d ± 5.55	6.68 b ± <0.001	2.60 d ± 0.05	3.35 c ± 0.12
Z	ND	ND	2.97 b ± 0.12	0.09 b ± 0.00	1100.33 c ± 26.36	9.47 a ± <0.001	10.95 a ± 0.04	10.46 a ± 0.08

* Values represent the mean ± standard error ($n_{\text{Simple sugars, Iso-α-acids}}$ = 2, $n_{\text{Salt, Total dissolved solids, Ethanol content}}$ = 3). ND: Non-detectable. Abbreviations of samples may be found in Table 1. Different letters within a column indicate that values are significantly different according to the least significant difference test (LSD; $p < 0.05$).

3.2. Consumer Sensory Evaluation and Biometrics

Table 5 shows the mean values and ANOVA results of the self-reported responses from the consumers' sensory tests. Significant differences ($p < 0.05$) between samples were observed for all attributes evaluated. In all samples, except for Z, the responses from overall liking were higher when rated at the end of the test after assessing each attribute, compared to the overall liking at the start (before assessing individual attributes). Spontaneous fermentation beers with raspberry (Framboise) and cherry (Kriek) flavors were the most liked overall (LF: 10.79; LK: 10.73) and also received the highest in bitterness (LF: 11.85; LK: 11.06), acidity (LF: 10.76; LK: 11.37) and aroma (LF: 9.50; LK: 9.53) liking scores. For sweetness liking, there were non-significant differences among the spontaneous (LK, LF) and bottom fermentation samples (C, H), but these were significantly different from the top fermentation beers (L, Z). On the other hand, C had the lowest liking of foam stability (6.79) compared to all other beers (10.20–11.14).

Table 5. Sensory acceptability (self-reported responses) of commercial beers.

Sample	Overall Liking-Start	Foam Stability	Foam Height	Bitter	Sweet	Acidity	Aroma	Overall Liking-End
LK	10.35 a* ± 0.56	10.20 a ± 0.38	8.31 b ± 0.38	11.06 a ± 0.42	9.53 a ± 0.46	11.37 a ± 0.50	9.53 a ± 0.44	10.73 a ± 0.50
LF	10.04 ab ± 0.47	11.14 a ± 0.52	11.16 a ± 0.54	11.85 a ± 0.46	9.56 a ± 0.52	10.76 a ±0.51	9.50 a ± 0.50	10.79 a ± 0.50
C	7.57 cd ± 0.55	6.79 b ± 0.52	6.28 c ± 0.59	8.92 bc ± 0.45	8.62 a ± 0.53	7.07 b ±0.51	7.54 bc ± 0.53	7.74 bc ±0.54
H	8.72 bc ± 0.49	10.58 a ± 0.39	10.60 a ± 0.40	8.76 bc ± 0.51	9.51 a ± 0.51	7.35 b ±0.47	8.31 ab ± 0.51	9.07 b ± 0.49
L	6.69 d ± 0.58	10.63 a ± 0.43	10.32 a ± 0.41	7.78 c ± 0.57	6.61 b ± 0.58	6.91 b ±0.52	7.03 bc ± 0.55	6.90 c ± 0.63
Z	7.65 cd ± 0.63	10.46 a ± 0.51	10.83 a ± 0.44	9.48 b ± 0.60	6.83 b ± 0.61	7.59 b ±0.59	6.73 c ± 0.57	7.34 c ± 0.65

* Values represent the mean ± standard error $N = 61$. Different letters within a column indicate that values are significantly different according to the least significant difference test (LSD; $p < 0.05$). Abbreviations of samples may be found in Table 1.

Figure 2 shows the principal components analysis for the sensory self-reported and emotional (biometric) responses, and chemical data. The principal component one (PC1) represented 49.40%, while PC2 accounted for 26.35% of data variability (Total = 75.75%). According to the factor loadings

(FL), descriptors such as relaxed 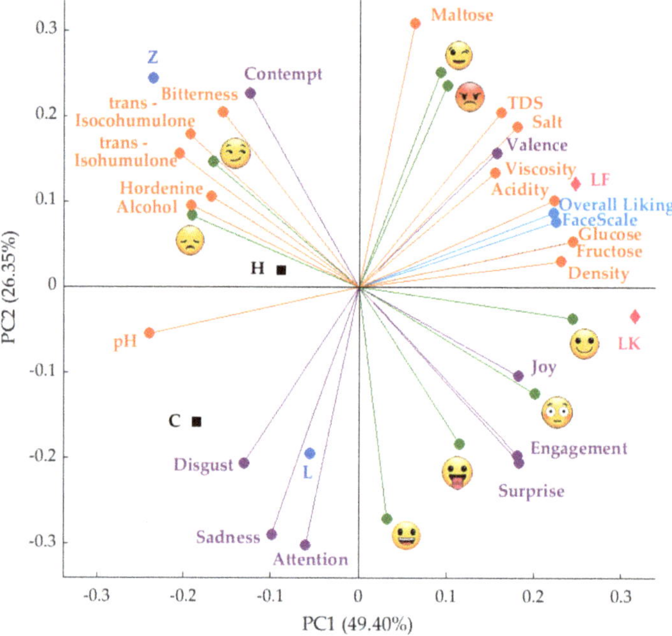 (FL = 0.24), glucose (FL = 0.24), fructose (FL = 0.24) and density (FL = 0.23) represented PC1 on the positive side of the axis; while pH (FL = −0.24), trans-Isohumulone (FL = −0.20) and trans-Isocohumulone (FL = −0.19) characterized it on the negative side. On the other hand, PC2 was represented by maltose (FL = 0.31), winking face (FL = 0.25) and rage (FL = 0.24) on the positive side; while attention (FL = −0.30), sadness (FL = −0.29), and smiley (FL = −0.27) represented it on the negative side of the axis. Sugars such as fructose and glucose were positively related to overall liking, FaceScale and relaxed, with the spontaneous fermentation beers (LK and LF) associated with those components. On the contrary, hordenine presented a negative relationship with the latter descriptors and a positive relationship with alcohol content, iso-alpha acids, bitterness, smirk , and disappointed , and beers such as H (bottom fermentation) and Z (top fermentation) were associated with these variables.

Figure 2. Principal components analysis showing the results from the sensory (self-reported liking responses in blue), biometrics (purple for emotions, and green for the emojis generated from the facial expressions), and chemical data (orange). Abbreviations: PC1 and PC2 = principal component 1 and 2, TDS = total dissolved solids. Abbreviations for samples are shown in Table 1.

Figure 3 shows the MFA for all chemicals, liking, and check all that apply data using emojis (Figure 3a) and emotion-terms (Figure 3b). In the MFA using emojis (Figure 3a), it can be observed that factors 1 and 2 (F1 and F2) represented 89.35% of total data variability (F1 = 68.86%; F2 = 20.49%). According to FL, the F1 was mainly represented by crying (FL = 1.13), pH (FL = 0.94), angry (FL = 0.88), and alcohol content (FL = 0.87) on the positive side of the axis, and by overall liking (FL = −0.99), FaceScale (FL = −0.99), glucose (FL = −0.94) and fructose (FL = −0.94) on the negative side. On the other hand, F2 was represented by crying (FL = 1.67), TDS (FL = 0.80), salt (FL = 0.74) and maltose (FL = 0.73) on the positive side, and by disappointed (FL = −0.50) and unamused (FL = −0.29) on the negative side of the axis. Hordenine was positively related to alcohol content, iso-alpha acids, bitterness, and emojis such as sick 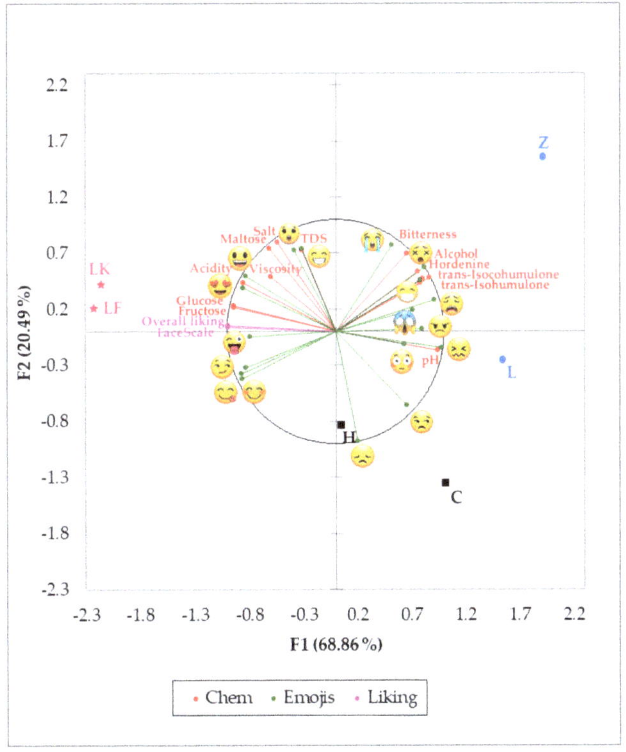, dizzy and weary , with top fermentation beer samples such as Z and L associated with those variables. In contrast, overall liking and FaceScale had a positive relationship with glucose, fructose, winking face with tongue , and love ; spontaneous fermentation samples LK and LF were most represented by these descriptors.

In Figure 3b, developed using emotion-terms, F1 and F2 accounted for 88.87% of total data variability (F1 = 70.50%; F2 = 18.37%). Based on the FL, variables such as FaceScale (FL = 0.99), overall liking (FL = 0.99), glucose (FL = 0.95) and fructose (FL = 0.95) represented the positive side of the F1 axis, while pH (FL = −0.95), acidity (FL = −0.88), aggressive (FL = −0.86), and alcohol content (FL = −0.86) characterized the negative side. Descriptors such as aggressive (FL = 0.84), TDS (FL = 0.77), maltose (FL = 0.72) and bitterness (FL = 0.71) represented the positive side of the F2 axis, whereas bored (FL = −0.54) and guilty (FL = −0.20) characterized the negative side. Similar to Figure 3a, hordenine had a positive relationship with alcohol content, iso-alpha acids, bitterness and emotion-terms such as aggressive, disgusted and nostalgic. These were negatively related with overall liking, FaceScale, fructose, glucose, acidity, joyful, affectionate, and happy. Samples were clearly grouped according to the type of fermentation: top (Z and L), bottom (C and H) and spontaneous (LF and LK).

(a)

Figure 3. *Cont.*

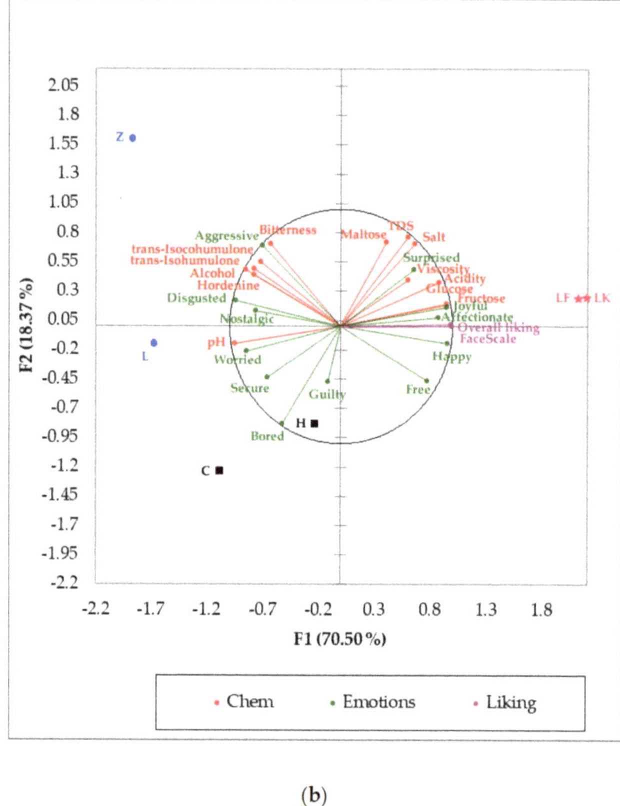

(b)

Figure 3. Multiple factor analysis showing the results from the sensory (self-reported liking responses), chemical data (Chem), and (**a**) emoji and (**b**) emotions, check all that apply responses. Abbreviations: F1 and F2 = Factor 1 and 2, TDS = total dissolved solids. Abbreviations for samples are shown in Table 1.

4. Discussion

Spontaneous fermentation beers resulted in the highest values for total sugars and lowest alcohol content, bitterness (expressed as IBU), and iso-alpha acids (Figure 1). This may be due to the addition of fruit juice (cherry in LK and raspberry in LF), and dried hops, which may also be old and oxidized to provide aromas and flavors but not bitterness [23,25,40].

There was a positive correlation ($R = 0.91$; $R^2 = 0.83$) between hordenine and alcohol content for all beer samples studied. The latter effect is in accordance with the study from Brauers et al. [16], who found higher hordenine content in strong beers (bock style), which have high alcohol content (6.6–7.5%; [41]), and lower hordenine values in alcohol-free beers. On the other hand, top fermentation beers were found to have higher concentrations of iso-alpha acids, hordenine, and bitterness (expressed as IBU) compared to the other samples (Figure 3). Sensorial bitterness can be derived from several compounds, including polyphenols and alkaloids [42].

For beers, 80% of the perceived bitterness is originated from adding hops during the brewing process [43]. Hops from female plants contain glands with a resin that is rich in derivates of phloroglucinol, essential oils, and flavonoids [44]. In terms of the bitter compounds, there are two types of acids in the hops resins, alpha, and beta; however, these molecules are not bitter in their raw forms. Before brewing, a thermal isomerization of the alpha-acids occurs during the boiling process, and iso-alpha acids are obtained, which are responsible for imparting the bitterness in beer. Two stereoisomers are generated during this isomerization process, trans- and cis-iso-alpha-acids, which are catalyzed by magnesium ions [45]. The perceived bitterness intensity is higher when there is a higher content of iso-alpha-acids. This compound provides a "harsh," "round," and "lingering" flavor to beer [43]. In the present study, the top fermentation beers (L and Z) had the lowest scores for the liking of bitterness compared to the other beer samples (Table 5). The higher chemical bitterness (expressed as IBU) for these two samples can potentially explain the disliking of the bitterness in the tasting session by the participants. Besides, hordenine is known to impart bitterness [18], and the concentration of this compound was also higher in the top fermentation beers.

Similar results were found using the conscious responses with emojis and words, and from the subconscious responses using biometrics. According to the PCA and MFA presented in Figures 2 and 3, respectively, beers with higher sugar content (glucose and fructose) were associated with positive emotions such as joy, relaxed 🙂, love 😍, winking face with tongue 😜, affectionate, and FaceScale in both subconscious and conscious responses (emojis and emotion-terms). This coincides with findings by Kim et al. [46], who reported that samples of beverages and biscuits with the highest sugar content elicited positive emotions such as affectionate, pleased, joyful, glad, and happy. On the other hand, bitterness has been associated with rejection due to genetic factors and the innate relationship of bitter products with poisonous compounds [24,47,48]. Overall taste liking is the result of the intrinsic balance among the basic tastes that are sensed by the receptors located in the gustative system [49]. Individual taste compounds can elicit discrete sensations in consumers. However, different tastes can interact with each other, which can result in suppression or enhancement effects of certain perceptions [50,51]. For instance, minor concentrations of sugar can enhance the sourness of citric acid solutions; or slight concentrations of salt can enhance the sweetness of sugar solutions. The opposite can also occur as slight concentrations of quinine (a bitter compound) mixed with saccharides can suppress the sweetness of the solutions [52,53]. This can potentially explain the overall taste perception by the consumers in the present study. As the sugar content of the spontaneous fermentation beers was higher compared to the other samples, the bitterness perception of those beers was somewhat suppressed, which produced higher hedonic and emotional responses. This effect can be observed for both responses (conscious and subconscious) measured in this study, as the sweet taste was the main factor responsible for the overall satisfaction of consumers.

Even though hordenine has been reported to stimulate the release of dopamine and is, therefore, associated with happiness [15,21], these studies have not evaluated these effects on consumers when drinking beer. In the present research, it was found that, as hordenine was positively related with bitterness and other bitter compounds such as iso-alpha acids, all these had a positive relationship with negative emotions such as disappointed 😖 (Figure 2), dizzy 😵, sick 🤢, weary 😩 (Figure 3a), disgusted, and aggressive (Figure 3b). This may be due to two main factors: (i) the higher sugar concentration in beers LF and LK, which had a higher effect on consumers, and (ii) the time of the sensory session, which may not have been long enough to increase hordenine concentration in the bloodstream significantly. Hence, since the effects of hordenine may be delayed, a sensory tasting session, including several sample beers, may not be appropriated to study the carry-over effects. This may be overcome by conducting further research allowing more time between beers for emotional assessments, so that there is enough hordenine level in the blood to more accurately assess the elicited emotional responses. Moreover, by comparing similar beer styles with alcoholic and non-alcoholic beers, it may render more information on the effects of hordenine and other compounds alone.

5. Conclusions

This preliminary study was a first attempt to associate beer compounds with the emotional responses of consumers using non-invasive biometrics. Findings showed that there was a positive relationship between sugar content, acidity, and positive emotions. At the same time, alcohol, bitterness, and hordenine were associated with negative emotions, which explain the consumers' preference for spontaneous fermentation samples, which are sweeter and less bitter than other beer styles. The strong correlation between alcohol and hordenine, along with the effect that time may have in terms of increasing the hordenine levels in the bloodstream, leads to the need to conduct further studies, which may allow giving more time between samples to assess emotional responses and to compare alcoholic and non-alcoholic beers with similar styles to separate the effects of alcohol and hordenine. Additionally, further studies may include the assessment of differences in emotional responses among consumers from different cultural backgrounds. Results from these studies may be useful for brewing companies to modify their products for different markets and satisfy the needs of distinct target consumers.

Author Contributions: Conceptualization, C.G.V., R.V.-L., D.D.T., C.H.-B., and S.F.; Data curation, C.G.V., and S.F.; Formal analysis, C.G.V., R.V.-L., Y.G.R.-V., Z.E.-A., R.M., C.H.-B. and S.F.; Funding acquisition, C.G.V., Z.E.-A., A.P.S., C.H.-B. and S.F.; Investigation, C.G.V., R.V.-L., D.D.T., C.H.-B. and S.F.; Methodology, C.G.V., R.V.-L., D.D.T., Y.G.R.-V., P.A.R.-P., and C.H.-B.; Project administration, C.G.V., and S.F.; Software, C.G.V., and S.F.; Supervision, C.G.V., Z.E.-A., A.P.S., C.H.-B., and S.F.; Validation, C.G.V., R.V.-L., Z.E.-A., P.A.R.-P, A.P.S., C.H.-B., and S.F.; Visualization, C.G.V., R.V.-L., D.D.T., C.H.-B., and S.F.; Writing—original draft, C.G.V., R.V.-L., D.D.T., C.H.-B., and S.F.; Writing—review & editing, C.G.V., R.V.-L., C.H.-B., and S.F. All authors have read and agreed to the published version of the manuscript.

Funding: Funding was provided by Universitas 21 Graduate Collaborative Research Awards, Tecnologico de Monterrey, Translational-Omics (GIEE) Research Group, and CONACYT-Mexican National Council for Research and Technology (Scholarships to Raul Villarreal-Lara no. 359813 and to Yaressi G. Rodríguez-Velazco no. 560895).

Acknowledgments: We are grateful to Jesus Salvador Jaramillo for his technical support with the bitterness spectrophotometric determinations, and to Dariana G. Rodríguez-Sánchez for her technical contributions to the iso-alpha acid quantifications.

Conflicts of Interest: The authors declare no conflict of interest.

References

1. Gonzalez Viejo, C.; Fuentes, S.; Li, G.; Collmann, R.; Condé, B.; Torrico, D. Development of a robotic pourer constructed with ubiquitous materials, open hardware and sensors to assess beer foam quality using computer vision and pattern recognition algorithms: RoboBEER. *Food Res. Int.* **2016**, *89*, 504–513. [CrossRef]
2. Bamforth, C.; Russell, I.; Stewart, G. *Beer: A Quality Perspective*; Academic Press: Burlington, MA, USA, 2011.
3. Gonzalez Viejo, C.; Fuentes, S.; Torrico, D.; Howell, K.; Dunshea, F. Assessment of beer quality based on foamability and chemical composition using computer vision algorithms, near infrared spectroscopy and machine learning algorithms. *J. Sci. Food Agric.* **2018**, *98*, 618–627. [CrossRef]
4. De Keukeleire, D. Fundamentals of beer and hop chemistry. *Quim. Nova* **2000**, *23*, 108–112. [CrossRef]
5. Bamforth, C. 125th Anniversary Review: The Non-Biological Instability of Beer. *J. Inst. Brew.* **2011**, *117*, 488–497. [CrossRef]
6. Briggs, D.E.; Hough, J.S.; Stevens, R.; Young, T.W. *Malting and Brewing Science: Hopped Wort and Beer*; Kluwer Academic/Plenum Publishers: St Edmundsbury, UK, 1982; Volume 2.
7. Pelletier, S.W. The nature and definition of an alkaloid. *Alkaloids: Chem. Biol. Perspect.* **1983**, *1*, 1–31.
8. Bart, R. *The Chemistry of Beer: The Science in the Suds*; Wiley: Hoboken, NJ, USA, 2013. [CrossRef]
9. Wong, S.; Lua, P. Chocolate: Food for moods. *Malays. J. Nutr.* **2011**, *17*, 259–269. [PubMed]
10. Mitchell, E.; Slettenaar, M.; Vd Meer, N.; Transler, C.; Jans, L.; Quadt, F.; Berry, M. Differential contributions of theobromine and caffeine on mood, psychomotor performance and blood pressure. *Physiol. Behav.* **2011**, *104*, 816–822. [CrossRef] [PubMed]
11. Duarte, A. Xylitol: Sweet and healthy. *Pac. J. Energy Med.* **2008**, *2*, 58–64.
12. Verna, R. The history and science of chocolate. *Malays. J. Pathol.* **2013**, *35*, 111.
13. Pietraszek, M.; Urano, T.; Sumioshi, K.; Serizawa, K.; Takahashi, S.; Takada, Y.; Takada, A. Alcohol-induced depression: Involvement of serotonin. *Alcohol Alcohol.* **1991**, *26*, 155–159. [CrossRef]
14. Duncan, M.; Smythe, G. Salsolinol and dopamine in alcoholic beverages. *Lancet* **1982**, *319*, 904–905. [CrossRef]
15. Sommer, T.; Hübner, H.; El Kerdawy, A.; Gmeiner, P.; Pischetsrieder, M.; Clark, T. Identification of the beer component hordenine as food-derived dopamine D2 receptor agonist by virtual screening a 3D compound database. *Sci. Rep.* **2017**, *7*, 44201. [CrossRef] [PubMed]

16. Brauers, G.; Steiner, I.; Daldrup, T. Quantification of the biogenic phenethylamine alkaloid hordenine by LC-MS/MS in beer. *Toxichem Krimtech* **2013**, *80*, 323–326.

17. Steiner, I.; Brauers, G.; Temme, O.; Daldrup, T. A sensitive method for the determination of hordenine in human serum by ESI+ UPLC-MS/MS for forensic toxicological applications. *Anal. Bioanal. Chem.* **2016**, *408*, 2285–2292. [CrossRef] [PubMed]

18. Shahidi, F.; Naczk, M. *Phenolics in Food and Nutraceuticals*; CRC Press: Boca Raton, FL, USA, 2003.

19. Perrine, D.M. Visions of the Night-Western Medicine Meets Peyote 1887–1899. *Heffter Rev. Psychedelic Res.* **2001**, *2*, 6–52.

20. Baixauli, E. Happiness: Role of Dopamine and Serotonin on mood and negative emotions. *Emerg. Med.* **2017**, *7*, 350. [CrossRef]

21. Oberlin, B.G.; Dzemidzic, M.; Tran, S.M.; Soeurt, C.M.; Albrecht, D.S.; Yoder, K.K.; Kareken, D.A. Beer flavor provokes striatal dopamine release in male drinkers: Mediation by family history of alcoholism. *Neuropsychopharmacology* **2013**, *38*, 1617–1624. [CrossRef]

22. Gonzalez Viejo, C.; Fuentes, S.; Howell, K.; Torrico, D.; Dunshea, F. Robotics and computer vision techniques combined with non-invasive consumer biometrics to assess quality traits from beer foamability using machine learning: A potential for artificial intelligence applications. *Food Control* **2018**. [CrossRef]

23. Gonzalez Viejo, C.; Fuentes, S.; Torrico, D.; Howell, K.; Dunshea, F. Assessment of Beer Quality Based on a Robotic Pourer, Computer Vision, and Machine Learning Algorithms Using Commercial Beers. *J. Food Sci.* **2018**, *83*, 1381–1388. [CrossRef]

24. Gonzalez Viejo, C.; Fuentes, S.; Howell, K.; Torrico, D.; Dunshea, F. Integration of non-invasive biometrics with sensory analysis techniques to assess acceptability of beer by consumers. *Physiol. Behav.* **2019**, *200*, 139–147. [CrossRef]

25. Gonzalez Viejo, C.; Fuentes, S.; Torrico, D.; Godbole, A.; Dunshea, F. Chemical characterization of aromas in beer and their effect on consumers liking. *Food Chem.* **2019**, *293*, 479–485. [CrossRef] [PubMed]

26. Piggott, J. *Alcoholic Beverages: Sensory Evaluation and Consumer Research*; Elsevier Science: Philadelphia, PA, USA, 2011.

27. Campbell-Platt, G. *Food Science and Technology*; Wiley: New York, NY, USA, 2017.

28. Pielech-Przybylska, K.; Balcerek, M.; Dziekońska-Kubczak, U.; Pacholczyk-Sienicka, B.; Ciepielowski, G.; Albrecht, Ł.; Patelski, P. The role of Saccharomyces cerevisiae yeast and lactic acid bacteria in the formation of 2-propanol from acetone during fermentation of rye mashes obtained using thermal-pressure method of starch liberation. *Molecules* **2019**, *24*, 610. [CrossRef] [PubMed]

29. Okafor, V.; Eboatu, A.; Anyalebechi, R.; Okafor, U. Comparative studies of the physicochemical properties of beers brewed with hop extracts and extracts from four selected tropical plants. *J. Adv. Chem. Sci.* **2016**, *2*, 382–386.

30. Heredia-Olea, E.; Cortés-Ceballos, E.; Serna-Saldívar, S.O. Malting Sorghum with Aspergillus Oryzae Enhances Gluten-Free Wort Yield and Extract. *J. Am. Soc. Brew. Chem.* **2017**, *75*, 116–121. [CrossRef]

31. Alonso-Gómez, L.A.; Heredia-Olea, E.; Serna-Saldivar, S.O.; Bello-Pérez, L.A. Whole unripe plantain (Musa paradisiaca L.) as raw material for bioethanol production. *J. Sci. Food Agric.* **2019**, *99*, 5784–5791.

32. Martin, R.; Collazo, D.; Davis, J.; Hamp, F.; Huestis, S.; Jaskula-Goiris, B.; Jordan, J.; Juzeler, R.; Krug, S.; Lee, K. Reduced hazardous solvent use for determination of isohumulone bitterness units. *J. Am. Soc. Brew. Chem.* **2011**, *69*, 276–277.

33. Vanhoenacker, G.; de Keukeleire, D.; Sandra, P. Analysis of iso-α-acids and reduced iso-α-acids in beer by direct injection and liquid chromatography with ultraviolet absorbance detection or with mass spectrometry. *J. Chromatogr. A* **2004**, *1035*, 53–61. [CrossRef] [PubMed]

34. Barber, L.; Arnberg, K.; Aron, P.; Mulqueen, S.; Raver, M.; Reffner, J.; Shellhammer, T.; Smith, B.; Porter, A. Determination of iso-α-acids in beer and wort by high-performance liquid chromatography. *J. Am. Soc. Brew. Chem.* **2011**, *69*, 288.

35. Sommer, T.; Dlugash, G.; Hübner, H.; Gmeiner, P.; Pischetsrieder, M. Monitoring of the dopamine D2 receptor agonists hordenine and N-methyltyramine during the brewing process and in commercial beer samples. *Food Chem.* **2019**, *276*, 745–753. [CrossRef]

36. Escobar, E.R.; Gamino, M.N.B.; Salazar, R.M.; Hernández, I.S.S.; Martínez, V.C.; Bautista, C.F.; López, M.d.L.G.; Buenabad, N.A.; Medina-Mora, M.E.; Velázquez, J.A.V. National trends in alcohol consumption in Mexico: Results of the National Survey on Drug, Alcohol and Tobacco Consumption 2016–2017. *Salud Ment.* **2018**, *41*, 7–15. [CrossRef]

37. Garcia, D. En México, los regios son los que más consumen cerveza. *Milenio* **2019**.

38. King, S.C.; Meiselman, H.L. Development of a method to measure consumer emotions associated with foods. *Food Qual. Prefer.* **2010**, *21*, 168–177. [CrossRef]

39. McDuff, D.; Mahmoud, A.; Mavadati, M.; Amr, M.; Turcot, J.; Kaliouby, R.E. AFFDEX SDK: A cross-platform real-time multi-face expression recognition toolkit. In Proceedings of the 2016 CHI Conference Extended Abstracts on Human Factors in Computing Systems, San Jose, CA, USA, 7–12 May 2016; pp. 3723–3726.

40. De Keersmaecker, J. The mystery of lambic beer. *Sci. Am.* **1996**, *275*, 74–80. [CrossRef]

41. Perozzi, C.; Beaune, H. *The Naked Pint: An Unadulterated Guide to Craft Beer*; Penguin Publishing Group: New York, NY, USA, 2012.

42. Wiener, A.; Shudler, M.; Levit, A.; Niv, M.Y. BitterDB: A database of bitter compounds. *Nucleic Acids Res.* **2012**, *40*, D413–D419. [CrossRef] [PubMed]

43. Oladokun, O.; Tarrega, A.; James, S.; Cowley, T.; Dehrmann, F.; Smart, K.; Cook, D.; Hort, J. Modification of perceived beer bitterness intensity, character and temporal profile by hop aroma extract. *Food Res. Int.* **2016**, *86*, 104–111. [CrossRef]

44. Alonso-Esteban, J.I.; Pinela, J.; Barros, L.; Ćirić, A.; Soković, M.; Calhelha, R.C.; Torija-Isasa, E.; de Cortes Sánchez-Mata, M.; Ferreira, I.C. Phenolic composition and antioxidant, antimicrobial and cytotoxic properties of hop (Humulus lupulus L.) Seeds. *Ind. Crops Prod.* **2019**, *134*, 154–159.

45. Haseleu, G.; Intelmann, D.; Hofmann, T. Structure determination and sensory evaluation of novel bitter compounds formed from β-acids of hop (Humulus lupulus L.) upon wort boiling. *Food Chem.* **2009**, *116*, 71–81. [CrossRef]

46. Kim, J.-Y.; Prescott, J.; Kim, K.-O. Emotional responses to sweet foods according to sweet liker status. *Food Qual. Prefer.* **2017**, *59*, 1–7. [CrossRef]

47. Dinehart, M.; Hayes, J.; Bartoshuk, L.; Lanier, S.; Duffy, V. Bitter taste markers explain variability in vegetable sweetness, bitterness, and intake. *Physiol. Behav.* **2006**, *87*, 304–313. [CrossRef] [PubMed]

48. Wang, X.; Thomas, S.D.; Zhang, J. Relaxation of selective constraint and loss of function in the evolution of human bitter taste receptor genes. *Hum. Mol. Genet.* **2004**, *13*, 2671–2678. [CrossRef]

49. Lawless, H.T.; Heymann, H. *Sensory Evaluation of Food: Principles and Practices*; Springer Science & Business Media: Berlin, Germany, 2010.

50. Keast, R.S.; Breslin, P.A. An overview of binary taste–taste interactions. *Food Qual. Prefer.* **2003**, *14*, 111–124. [CrossRef]

51. Torrico, D.D.; Prinyawiwatkul, W. Psychophysical effects of increasing oil concentrations on saltiness and bitterness perception of oil-in-water emulsions. *J. Food Sci.* **2015**, *80*, S1885–S1892. [CrossRef] [PubMed]

52. Kiefer, S.W.; Lawrence, G.J. The sweet–bitter taste of alcohol: Aversion generalization to various sweet–quinine mixtures in the rat. *Chem. Senses* **1988**, *13*, 633–641. [CrossRef]

53. Bartoshuk, L.; Cleveland, C. Mixtures of substances with similar tastes. A test of a psychophysical model of taste mixture interactions. *Sens. Process.* **1977**, *1*, 177.

foods

Article

Differences in Hedonic Responses, Facial Expressions and Self-Reported Emotions of Consumers Using Commercial Yogurts: A Cross-Cultural Study

Mitali Gupta [1,2,*], Damir D. Torrico [3], Graham Hepworth [4], Sally L. Gras [2,5], Lydia Ong [2,5], Jeremy J. Cottrell [1,2] and Frank R. Dunshea [1,2,6]

[1] School of Agriculture and Food, Faculty of Veterinary and Agricultural Sciences, The University of Melbourne, Parkville, VIC 3010, Australia; jcottrell@unimelb.edu.au (J.J.C.); fdunshea@unimelb.edu.au (F.R.D.)

[2] Future Food Hallmark Research Initiative Project, The University of Melbourne, Parkville, VIC 3010, Australia; sgras@unimelb.edu.au (S.L.G.); lon@unimelb.edu.au (L.O.)

[3] Department of Wine, Food and Molecular Biosciences, Lincoln University, Lincoln 7647, New Zealand; Damir.Torrico@lincoln.ac.nz

[4] Statistical Consulting Centre, The University of Melbourne, Melbourne, VIC 3010, Australia; hepworth@unimelb.edu.au

[5] Department of Chemical Engineering and The Bio21 Molecular Science and Biotechnology Institute, The University of Melbourne, Parkville, VIC 3010, Australia

[6] Faculty of Biological Sciences, The University of Leeds, Leeds LS2 9JT, UK

* Correspondence: mitalik@student.unimelb.edu.au; Tel.: +61-3-8344-1854

Citation: Gupta, M.; Torrico, D.D.; Hepworth, G.; Gras, S.L.; Ong, L.; Cottrell, J.J.; Dunshea, F.R. Differences in Hedonic Responses, Facial Expressions and Self-Reported Emotions of Consumers Using Commercial Yogurts: A Cross-Cultural Study. *Foods* **2021**, *10*, 1237. https://doi.org/10.3390/foods10061237

Academic Editor: Han-Seok Seo

Received: 12 April 2021
Accepted: 25 May 2021
Published: 29 May 2021

Publisher's Note: MDPI stays neutral with regard to jurisdictional claims in published maps and institutional affiliations.

Abstract: Hedonic scale testing is a well-accepted methodology for assessing consumer perceptions but is compromised by variation in voluntary responses between cultures. Check-all-that-apply (CATA) methods using emotion terms or emojis and facial expression recognition (FER) are emerging as more powerful tools for consumer sensory testing as they may offer improved assessment of voluntary and involuntary responses, respectively. Therefore, this experiment compared traditional hedonic scale responses for overall liking to (1) CATA emotions, (2) CATA emojis and (3) FER. The experiment measured voluntary and involuntary responses from 62 participants of Asian (53%) versus Western (47%) origin, who consumed six divergent yogurt formulations (Greek, drinkable, soy, coconut, berry, cookies). The hedonic scales could discriminate between yogurt formulations but could not distinguish between responses across the cultural groups. Aversive responses to formulations were the easiest to characterize for all methods; the hedonic scale was the only method that could not characterize differences in cultural preferences, with CATA emojis displaying the highest level of discrimination. In conclusion, CATA methods, particularly the use of emojis, showed improved characterization of cross-cultural preferences of yogurt formulations compared to hedonic scales and FER.

Keywords: biometrics; Cochran's Q test; ethnic; plant; conscious; unconscious; check-all-that-apply; linear model; correspondence analysis

1. Introduction

Sensory analysis is important in the food industry, not only for product development but also to guide marketing decisions [1]. While simple sensory testing using a traditional hedonic scale has been widely used to understand the acceptance of foods, this approach has limited freedom for the expression of a full range of sensory experiences [2]. Furthermore, taste responsiveness differs by ethnicity and gender, which may generate different perceptions when tasting the same product. However, these differences may not be reflected when scoring by a hedonic scale and can generate similar scores [3]. Hence, it is important to understand the affective responses of consumers and how people of different cultures perceive food products, especially in a multicultural environment, using a more elaborate approach [4].

A hedonic scale is a simple method of measuring the overall liking scores, which is considered the most informative assessment by a consumer. However, consumer perception of a product goes beyond overall liking, since it also includes extrinsic elements such as the perceived benefits, quality and wellness of the associated food. These extrinsic elements can be explained using more advanced approaches, such as the EsSense Profile™ [5] and check-all-that-apply (CATA)—conscious methods—and the recognition of the facial expression—an unconscious method [6]. The CATA methods can be based on emotion terms or emojis [7,8]. In previous studies, the CATA methodology has been effectively used to compare consumer perception and liking of different chocolate milk desserts and beers and has been shown to be an easy and convenient method for understanding consumer behaviors [9,10]. This method has also been used to observe cultural differences in terms of word associations with the relative quality criteria of rice consumption [11]. An easy, non-verbal CATA method is the use of emoji and emoticons to express emotions towards the different food types [12]. Emoji questionnaires are suitable for a range of populations, as people find emojis to be more expressive compared to emotional terms [7,11]. However, there can be drawbacks with this technique, as it may not detect subtle differences between sensory characteristics [13]. Moreover, this technique is still considered a self-reported conscious method, because it asks consumers to select options from a list and therefore can show bias.

Novel techniques for the assessment of consumer liking involve recording the unconscious facial expressions of consumers during product tasting. This approach can provide details of unconscious emotions expressed by consumers, rather than asking consumers to select emotional terms. Unconscious consumer responses can be measured using facial expression recognition technology, such as FaceReader™ software. This instrumental analysis measures unconscious consumer emotions by reading the face and classifying facial expressions while tasting [14]. It is an easy and quick technique to understand the emotions expressed towards tasted food products. The technique can address any bias, as it records the emotions of the consumers intrinsically [15], although it does not record liking. Facial expression recognition technologies are promising for increasing our understanding of the link between consumer food choices and the facial expressions induced by different the food tastes [16]. In a study, FaceReader™ was used to assess the instant facial expressions of students with an efficacy of 87%. Students were found to experience neutral, sad and angry emotions during assessment [17]. However, this technology can generate noise within the data, which can give false assessments [18].

Benefits may potentially be realized by integrating the different sensory procedures available. Specifically, this may allow a better understanding of consumer perceptions for different products in a multicultural environment. Previous studies have shown a significant effect of culture on the liking and consumer acceptance of a yogurt product and on the form or inclusions preferred by each of the ethnic groups. This makes it significantly important to understand the role or acceptance of the yogurts by different cultures using a cross-cultural sensory experiment [19]. There have been several studies involving various products, such as coffee labels [20], chocolate [21], beer [22] and artificial sweeteners [23], where a combination of implicit and explicit sensory methods have been successfully integrated together to assess the consumer acceptability. In the case of the study with coffee labels, the biometrics technology of facial expression recognition was also shown to be a reliable method for sensory evaluation [20]. However, the sensory analysis of chocolate using other biometrics responses, such as skin temperature and heart rate, did not show any significant differences for the tasted samples [21].

In the present study, dairy yogurts and their plant alternatives were taken as a reference for comparison of the different sensory techniques. Previously, plant-based yogurt alternatives have been shown to be perceived differently by consumers, as compared to dairy yogurts [24], which could produce different emotional experiences. The objective of this study was to understand the effect of culture on the conscious (CATA method using emotions and emojis) and unconscious (facial expression recognition) emotional responses. Further, the study sought to compare these methods with a traditional hedonic scale to understand the method which best represents liking towards yogurts. The study also sought to understand the purchase intent and willingness to pay for these yogurt types. The paper is sub-divided into mainly three sub-sections: (1) measuring liking for the yogurts using each of the four methods (hedonic scales, CATA emotions, CATA emojis and FER) individually and comparing across the Asian and Western cultures, (2) comparing and finding relationships between the methods using a linear mixed model and multi-factor analysis and (3) understanding price perception and purchase intent of the tasted yogurts.

2. Materials and Methods

The participants were recruited for the tasting sessions through email invitations and all protocols of this study were approved by the Human Ethics Advisory Group (HEAG) of the Faculty of Veterinary and Agriculture Sciences (FVAS) at the University of Melbourne (Ethics ID 1853507.2). Consumers allergic to lactose, or those with any other allergies, such as wheat or nuts, were excluded from the study. Participants were provided with a gift voucher as an incentive for participation.

2.1. Samples

Six yogurt samples that differed in terms of taste and texture were selected, so that it was easier to capture the different consumer responses and understand the method that best relates to consumer liking and expectation. Plant-based yogurts are a new category of products, and recent reports are available mentioning that these are not as acceptable as their dairy counterparts [25]. The present study aimed to compare a traditional dairy product with recently popular plant-based alternatives. Furthermore, comparing different forms (drinkable) and inclusions (berry or cookies) in yogurt also might affect the consumer acceptability. Using a focus group study ($n = 32$), the yogurts were selected for the main tasting experiment, which included the popular options available commercially, and also for the selection of emotion terms in the experiment. The product codes, listed in Table 1, represent the following yogurt types: dairy Greek plain (reference), coconut plain (plant), soy plain (plant), drinkable (sweetened), dairy with crunchies (sweetened) and dairy with berry (sweetened).

Table 1. Comparison of overall liking scores of the different yogurt products by Asian and Western consumers.

Yogurt Product Type	Product Code (as Used in This Study)	Overall Liking Scores (Asian Consumers)	Overall Liking Scores (Western Consumers)
Dairy Greek yogurt (plain)	Reference	5.03 ± 2.33 [cd]	5.39 ± 2.26 [bc]
Coconut-based yogurt (plain)	Coconut	3.73 ± 2.05 [d]	4.39 ± 2.30 [c]
Drinkable yogurt (sweetened)	Drinkable	6.15 ± 2.16 [b]	5.51 ± 2.11 [b]
Soy-based yogurt (plain)	Soy	5.45 ± 2.25 [bc]	5.64 ± 2.11 [b]
Dairy yogurt with crunchies (sweetened)	Cookies	7.82 ± 0.94 [a]	6.87 ± 1.43 [a]
Dairy yogurt with berries (sweetened)	Berry	1.90 ± 0.92 [e]	2.02 ± 1.12 [d]
* F-value		38.93	20.86

[a,b,c,d,e] Means with different superscripts in each column indicate significant differences ($p < 0.05$) by Fisher's least square difference test. Highest value with 'a' in superscript. The data are presented as mean ± standard deviation. * as an indicator of significance.

2.2. Participants

A total of $n = 62$ participants answered the questions for the sensory tasting of the six different yogurt samples. The participants included 47% self-identified Western consumers (29) and 53% Asian consumers (33), 68% females (42) and 32% males (20), with ages ranging from 21 to 58 years. The Western consumers were mainly Australian (14), European (8),

North/South American (3) and Latin (4), and the Asian consumers were mainly Chinese (19), Indian/Sri Lankan/Bangladeshi (7), Korean (2), Filipino (1), Persian (1), Vietnamese (2) and Malaysian (1). All participants were students or staff members at the University of Melbourne.

2.3. Sensory Evaluation and Data Collection

Participants were asked to taste each of the six yogurt samples (~15 g), which were served in plastic containers at an internal temperature of $10 \pm 2\,°C$, labeled with 3-digit codes and presented in a random order. Participants were seated in individual sensory booths (under white natural LED light); the temperature of the booths was maintained at $22 \pm 2\,°C$. Each booth was equipped with a Samsung 18-inch tablet (Samsung Group, Seoul, South Korea) for recording videos to obtain consumer responses using the Bio-sensory app [26], which did not require any calibration. The armed tablets were placed near to face of the participants, with an in-built camera, and could be adjusted according to the participant's height. When a participant was seated in the sensory booth, it was ensured that their faces were recorded properly in the set-up, at an approximate distance of 45–50 cm from the participant. The investigators who had expertise in sensory ensured at the beginning of the tasting that the participants' faces were recorded properly in the set-up.

There was a mandatory break of at least 30 s in between the tasting of any two samples for cleansing of the palette with crackers and water. For each product, participants were asked to evaluate overall liking using a continuous 9-point hedonic scale (1—dislike extremely, 5—neither dislike nor like and 9—like extremely). The self-reported emotional responses of participants were recorded using the CATA methodology, where they were asked to choose the terms from a list of emotions provided (Table S1). These emotion terms were divided into positive, negative and neutral categories and were selected using the perceptual mapping technique [27,28]. A similar procedure was implemented for choosing emojis from another list (Table S2), which were representations of popular face scales from other tasting studies [7]. Preliminary research group discussions also helped in reaching the final list of terms of emotions and emojis, which were fitting for the cross-cultural component of the study.

Facial expressions were recorded during the tasting of each yogurt (Table S3). The videos were cut from the point when participants placed the spoonful of yogurt in their mouth until the time it was swallowed (or consumed), showing their first immediate reaction, which was covered in 5–7 s or even less time in most cases. Further, the data were processed using FaceReader^{TM} 8.0 software (Noldus Information Technology, Wageningen, Netherlands) for recording facial emotions. No calibration was required for the FaceReader^{TM} software as it has pre-built models to detect faces and analyze the expressions. The video analysis was carried out using the "general" settings for all the participants, except for the participants from South East Asia, for whom specific settings were used for analysis. Participants were also asked to assess purchase intent and willingness to pay for each product.

2.4. Data Analysis

Overall liking for each sample on the hedonic scale, purchase intent and willingness to pay were analyzed with ANOVA, and Fisher's LSD was used for paired comparisons, with a significance level of 0.05. The CATA analysis for emotions and emojis was performed with the Cochran's Q test and correspondence analysis using XLSTAT (Addinsoft, New York, NY, USA: version 2020.1.1). Correspondence analysis was carried out by the software using the variables generated in the Cochran's Q test. Fisher's exact test was applied to 2×2 cross-tables for comparison of the terms in CATA analysis between the two cultures. ANOVA was carried out on biometric emotions, and the PCA biplots were constructed using XLSTAT (Addinsoft, New York, NY, USA: version 2020.1.1). The linear mixed model for each method was created using REML in GENSTAT (VSN International, Hemel Hempstead, UK; version 16), after selecting the top descriptors using all-subsets regression

($p < 0.05$). Overall liking scores generated from hedonic liking were taken as response variables, and the emotion/emoji/biometric variables were taken as fixed factors. The random factors in the mixed model were "participant codes" and "product codes". If the fitting of the model did not converge numerically, one or both random factors were omitted to obtain convergence. A multi-factor analysis (MFA) was carried out for a combination of CATA emotions, CATA emojis, FER and overall liking using XLSTAT, with the mean values for each of the factors.

3. Results

3.1. Overall Liking of Yogurt Products

The mean overall liking scores for yogurt products for both Asian and Western participants were higher for the crunchy yogurt and lower for the berry yogurt (Table 1). The liking for the other four yogurt types was in between these two products, with the drinkable yogurt showing a higher liking by Asian consumers compared to the reference yogurt, and the coconut yogurt showing a lower liking compared to the drinkable and soy yogurts. In contrast, Western consumers rated these three samples (soy, drinkable and reference) similarly, with coconut having a lower liking. The interaction of products and culture was not significant ($p > 0.05$); hence, the differences between yogurt samples could not be observed across the two cultures.

3.2. Comparison of Emotion Terms (CATA Emotions Method)

The emotional terms expressed by consumers of Western and Asian backgrounds were compared by the Cochran's Q test (Table 2). The most selected emotional terms ($p < 0.05$) for Western consumers for each product type were 'trusted' and 'dependable' for the reference, 'artificial' and 'luxury' for coconut, 'cheerful' and 'uplifting' for drinkable, 'cheerful' and 'artificial' for soy, 'cheerful' and 'uplifting' for cookies and 'nasty' and 'deceitful' for berry. The distributions of positive, negative and neutral terms are listed in Table S1. In comparison, for Asian consumers, the top selected emotions ($p < 0.05$) for each of the yogurts were 'basic' and 'common' for the reference, 'artificial' for coconut, 'cheerful' and 'basic' for drinkable, 'cheerful' and 'artificial' for soy, 'cheerful' and 'luxury' for cookies and 'nasty' and 'artificial' for berry. The cookies product generated positive emotion terms, whereas berry generated negative emotions, by both the cultural groups. The emotion terms 'neutral' and 'guilt-free' were similarly rated by both cultures for all the yogurt products.

Interestingly, the reference product was related more to positive terms by Western consumers and to neutral terms by Asian consumers. The coconut was linked to more negative emotions by Asian consumers but was intermediate for Western consumers. The term 'common' was not significantly different for the six yogurt products in the case of Western consumers, whereas Asians selected 'common' more often for the reference yogurt, as seen by the Fisher's exact test across cultures.

A correspondence analysis using a biplot for Western consumers (Figure 1a) indicated that 82.47% of the variability observed could be attributed to the first two principal components (F1: 62.29%, F2: 20.18%). The berry was more related to the negative terms (nasty, pretentious, deceitful), whereas the reference (guilt-free, trusted, dependable) and soy (luxury, uplifting, cheerful, artificial) were more related to positive emotions. The drinkable and cookies yogurts were related to more positive and neutral terms ('common', 'basic', 'indifferent', 'luxury', 'cheerful', 'uplifting') and coconut was associated with neutral and negative responses ('cheap', 'neutral'). The biplot for Asian consumers (Figure 1b) similarly indicated that 85.5% of the variability observed could be attributed to the two principal components for (F1: 63.67%, F2: 21.49%). The berry and coconut yogurts were related to more negative emotion terms ('nasty', 'artificial', 'cheap', 'deceitful'); the reference dairy yogurt was related to neutral emotions ('guilt-free', 'common', 'neutral') and the cookies yogurt was more associated with positive emotions ('cheerful', 'luxury', 'uplifting'). The drinkable and soy yogurts were very similar to each other.

Table 2. Cochran's Q test for emotion terms shown for the two cultural groups (Western consumers, Asian consumers) comparing terms within the culture, and using Fisher's test to compare terms across cultures.

	Western Consumers							Asian Consumers						
Attributes	Reference	Coconut	Drinkable	Soy	Cookies	Berry	*p*-Values	Attributes	Reference	Coconut	Drinkable	Soy	Cookies	Berry	*p*-Values
Cheerful **	0.17 [bcd]	0.14 [cd]	0.38 [abc]	0.52 [ab]	0.59 [a]	0.00 [d]	<0.001	Cheerful **	0.12 [xy]	0.09 [xy]	0.36 [y]	0.39 [y]	0.88 [z]	0.00 [x]	<0.001
Neutral [ns]	0.21 [a]	0.24 [a]	0.35 [a]	0.17 [a]	0.14 [a]	0.17 [a]	0.280	Neutral [ns]	0.24 [x]	0.21 [x]	0.18 [x]	0.12 [x]	0.09 [x]	0.09 [x]	0.341
Nasty **	0.10 [b]	0.21 [b]	0.03 [b]	0.07 [b]	0.00 [b]	0.72 [a]	<0.001	Nasty **	0.15 [xy]	0.21 [xy]	0.09 [x]	0.09 [x]	0.00 [x]	0.55 [y]	<0.001
Luxury *	0.07 [b]	0.24 [ab]	0.10 [b]	0.14 [ab]	0.38 [a]	0.00 [b]	0.001	Luxury **	0.06 [x]	0.21 [xy]	0.03 [x]	0.18 [xy]	0.33 [y]	0.00 [x]	0.000
Guilt-free [ns]	0.24 [a]	0.14 [a]	0.10 [a]	0.07 [a]	0.17 [a]	0.00 [a]	0.077	Guilt-free *	0.24 [x]	0.09 [x]	0.09 [x]	0.03 [x]	0.03 [x]	0.09 [x]	0.045
Deceitful *	0.00 [b]	0.10 [ab]	0.07 [ab]	0.10 [ab]	0.03 [b]	0.28 [a]	0.003	Deceitful [ns]	0.09 [x]	0.09 [x]	0.12 [x]	0.15 [x]	0.00 [x]	0.12 [x]	0.313
Trusted **	0.34 [a]	0.03 [b]	0.14 [ab]	0.07 [b]	0.24 [ab]	0.00 [b]	0.000	Trusted *	0.24 [y]	0.03 [xy]	0.09 [xy]	0.15 [xy]	0.18 [xy]	0.00 [x]	0.007
Basic [ns]	0.24 [a]	0.24 [a]	0.31 [a]	0.21 [a]	0.10 [a]	0.10 [a]	0.193	Basic *	0.36 [y]	0.18 [xy]	0.33 [y]	0.15 [xy]	0.24 [xy]	0.03 [x]	0.008
Pretentious [ns]	0.00 [a]	0.10 [a]	0.00 [a]	0.00 [a]	0.00 [a]	0.07 [a]	0.060	Pretentious [ns]	0.06 [x]	0.12 [x]	0.00 [x]	0.09 [x]	0.00 [x]	0.06 [x]	0.094
Uplifting **	0.10 [bc]	0.20 [abc]	0.28 [abc]	0.31 [ab]	0.45 [a]	0.00 [c]	0.000	Uplifting **	0.06 [x]	0.09 [x]	0.24 [xy]	0.15 [x]	0.51 [y]	0.00 [x]	<0.001
Indifferent [ns]	0.14 [a]	0.21 [a]	**0.14 [a]**	0.03 [a]	0.03 [a]	0.07 [a]	0.210	Indifferent [ns]	0.06 [x]	0.09 [x]	**0.00 [x]**	0.09 [x]	0.03 [x]	0.12 [x]	0.315
Cheap [ns]	0.07 [a]	0.03 [a]	0.17 [a]	0.14 [a]	0.10 [a]	0.14 [a]	0.562	Cheap *	0.06 [x]	0.21 [x]	0.27 [x]	0.18 [x]	0.06 [x]	0.33 [x]	0.015
Dependable *	0.24 [a]	0.03 [ab]	0.07 [ab]	0.00 [b]	0.17 [ab]	0.00 [b]	0.002	Dependable [ns]	0.09 [x]	0.00 [x]	0.06 [x]	0.06 [x]	0.09 [x]	0.00 [x]	0.247
Artificial *	0.10 [b]	0.38 [ab]	0.24 [ab]	0.48 [a]	0.24 [ab]	0.24 [ab]	0.023	Artificial **	0.18 [xy]	0.52 [z]	0.27 [xyz]	0.36 [xyz]	0.06 [x]	0.46 [yz]	0.000
Common [ns]	**0.07 [a]**	0.10 [a]	0.10 [a]	0.07 [a]	0.10 [a]	0.03 [a]	0.911	Common *	**0.36 [y]**	0.09 [x]	0.15 [xy]	0.21 [xy]	0.15 [xy]	0.06 [x]	0.013

** Indicates significant differences between samples according to Cochran's Q test at *p* < 0.001. * Indicates significant differences between samples according to Cochran's Q test at *p* < 0.05. [ns] Indicates no significant difference between samples according to Cochran's Q test at *p* > 0.05, with 'a' the largest in case of Western and 'x' in case of Asian consumers. Significant differences between pairs of samples across cultures according to Fisher's exact test are indicated by bold terms. Fisher's exact test is generally more conservative than Cochran's Q test, so the paired comparisons may all be not significant even if the overall test is significant.

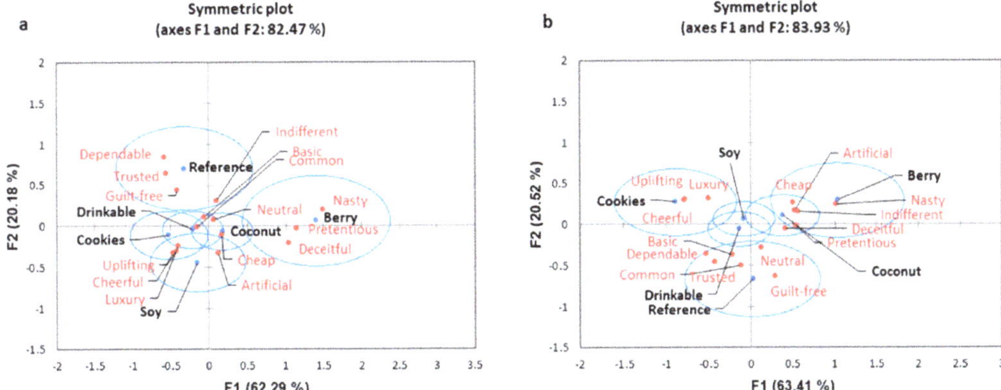

Figure 1. Correspondence analysis biplots for emotions comparing the two cultural groups: (**a**) Western consumers, (**b**) Asian consumers, with the confidence ellipses showing emotions related to each of the yogurt products.

3.3. Comparison of Emojis (CATA Emojis Method)

The emojis expressed by consumers of Western and Asian groups were assessed by the Cochran's Q test (Table 3). The most selected emotion terms ($p < 0.05$) for Western consumers were: 'smiling face' and 'relieved face' (reference); 'smiling face', 'persevering face' and 'savoring delicious food' (coconut); 'relieved face', 'smiling face' and 'savoring delicious food' (drinkable); 'relieved face' and 'smiling face' (soy); 'smiling face' and 'relieved face' (cookies); 'pensive face' (berry). In comparison to Asian consumers, the top selected emotions ($p < 0.05$) for the yogurts were: 'smiling face' and 'relieved face' (reference); 'relieved face' (coconut); 'relieved face' and 'smiling face' (drinkable); 'smiling face' (soy); 'smiling face' and 'savoring delicious food' (cookies); 'fearful' (berry). The reference, drinkable, soy and cookies yogurts were related to positive emojis, whereas berry was related to negative emojis, by both the cultural groups. The emojis 'unamused' and 'disappointed' were similarly rated by both cultures for all the yogurt products. The berry yogurt was chosen to be more 'pensive face' by Western consumers as compared to the Asians, and the cookies yogurt was related to 'relieved face' more often by Western consumers as compared to Asians, as seen by the Fisher's exact test across cultures.

The correspondence analysis biplot for Western consumers (Figure 2a) indicates that 86.93% of the variability observed could be attributed to the first two principal components (F1: 76.42%, F2: 10.51%). Meanwhile, a biplot for Asian consumers (Figure 2b) found that 89.76% variability (F1: 68.94%, F2: 20.83%) could be attributed to these components. Both yogurt products, coconut and berry, were related to more negative emojis by Asian as well as Western consumers, and the cookies was placed in the positive region by both cultural groups. In the case of Western participants, the reference, drinkable and soy yogurts were placed in between the positive and negative emotional areas, whereas Asian participants related the drinkable with positive emojis, and the reference and soy were placed in between the positive and negative regions.

Table 3. Cochran's Q test for emoji terms shown for the two cultural groups (Western consumers, Asian consumers) comparing terms within the culture, and using Fisher's test to compare terms across cultures.

Western Consumers								Asian Consumers							
Attributes	Reference	Coconut	Drinkable	Soy	Cookies	Berry	*p*-Values	Attributes	Reference	Coconut	Drinkable	Soy	Cookies	Berry	*p*-Values
[emoji] ns	0.14 ^{a}	0.21 ^{a}	0.21 ^{a}	0.10 ^{a}	**0.14** ^{a}	0.14 ^{a}	0.817	[emoji] ns	0.12 ^{xy}	0.24 ^{x}	0.12 ^{xy}	0.15 ^{xy}	**0.00** ^{y}	0.09 ^{xy}	0.087
[emoji] *	0.28 ^{ab}	0.21 ^{ab}	0.24 ^{ab}	0.31 ^{ab}	0.41 ^{a}	0.00 ^{b}	0.007	[emoji] **	0.15 ^{yz}	0.09 ^{z}	0.42 ^{xy}	0.30 ^{xyz}	0.52 ^{x}	0.00 ^{z}	<0.001
[emoji] ns	0.03 ^{a}	0.03 ^{a}	0.03 ^{a}	0.00 ^{a}	0.00 ^{a}	0.10 ^{a}	0.267	[emoji] *	0.00 ^{y}	0.09 ^{xy}	0.00 ^{y}	0.03 ^{xy}	0.00 ^{y}	0.18 ^{x}	0.001
[emoji] *	0.07 ^{b}	0.14 ^{ab}	0.03 ^{b}	0.07 ^{b}	0.03 ^{b}	**0.35** ^{a}	0.001	[emoji] ns	0.06 ^{x}	0.15 ^{x}	0.06 ^{x}	0.09 ^{x}	0.00 ^{x}	**0.12** ^{x}	0.255
[emoji] ns	0.10 ^{a}	0.00 ^{a}	0.00 ^{a}	0.03 ^{a}	0.10 ^{a}	0.00 ^{a}	0.098	[emoji] *	0.03 ^{xy}	0.00 ^{y}	0.03 ^{xy}	0.03 ^{xy}	0.18 ^{x}	0.00 ^{y}	0.004
[emoji] **	0.10 ^{ab}	0.10 ^{ab}	0.00 ^{b}	0.03 ^{b}	0.00 ^{b}	0.31 ^{a}	0.000	[emoji] *	0.21 ^{x}	0.09 ^{x}	0.03 ^{x}	0.12 ^{x}	0.00 ^{x}	0.21 ^{x}	0.026
[emoji] ns	0.03 ^{a}	**0.21** ^{a}	0.21 ^{a}	0.17 ^{a}	0.17 ^{a}	0.00 ^{a}	0.051	[emoji] *	0.09 ^{xy}	**0.00** ^{y}	0.06 ^{y}	0.09 ^{xy}	0.27 ^{x}	0.00 ^{y}	0.001
[emoji] ns	0.28 ^{a}	0.14 ^{a}	0.24 ^{a}	0.35 ^{a}	0.07 ^{a}	0.10 ^{a}	0.050	[emoji] ns	0.15 ^{x}	0.18 ^{x}	0.09 ^{x}	0.18 ^{x}	0.03 ^{x}	0.03 ^{x}	0.146
[emoji] **	0.00 ^{b}	0.17 ^{ab}	0.03 ^{b}	0.14 ^{ab}	0.00 ^{b}	0.31 ^{a}	0.000	[emoji] **	0.06 ^{y}	0.15 ^{xy}	0.00 ^{y}	0.09 ^{y}	0.00 ^{y}	0.33 ^{x}	<0.001
[emoji] *	0.10 ^{ab}	0.17 ^{ab}	0.24 ^{ab}	0.21 ^{ab}	0.35 ^{a}	0.00 ^{b}	0.005	[emoji] **	0.06 ^{y}	0.03 ^{y}	0.15 ^{y}	0.09 ^{xy}	0.42 ^{x}	0.00 ^{y}	<0.001
[emoji] **	0.07 ^{b}	0.07 ^{b}	0.00 ^{b}	0.00 ^{b}	0.00 ^{b}	0.31 ^{a}	<0.001	[emoji] **	0.06 ^{y}	0.09 ^{y}	0.06 ^{y}	0.06 ^{y}	0.00 ^{y}	0.49 ^{x}	<0.001
[emoji] *	0.00	0.00	0.03 ^{ab}	0.03 ^{ab}	0.00	0.17 ^{b}	0.004	[emoji] *	0.06 ^{xy}	0.03 ^{y}	0.03 ^{y}	0.00 ^{y}	0.00 ^{y}	0.21 ^{x}	0.001
[emoji] **	0.28 ^{ab}	0.07 ^{b}	0.45 ^{a}	0.35 ^{ab}	**0.55** ^{a}	0.03 ^{b}	<0.001	[emoji] *	0.18 ^{xy}	0.18 ^{xy}	0.30 ^{x}	0.12 ^{xy}	**0.15** ^{xy}	0.00 ^{y}	0.014

** Indicates significant differences between samples according to Cochran's Q test at *p* < 0.001. * Indicates significant differences between samples according to Cochran's Q test at *p* < 0.05. ns Indicates no significant difference between samples according to Cochran's Q test at *p* > 0.05, with 'a' the largest in case of Western and 'x' in case of Asian consumers. Significant differences between pairs of samples according to Fisher's exact test are indicated by bold terms. Fisher's exact test is generally more conservative than Cochran's Q test, so the paired comparisons may all be not significant even if the overall test is significant.

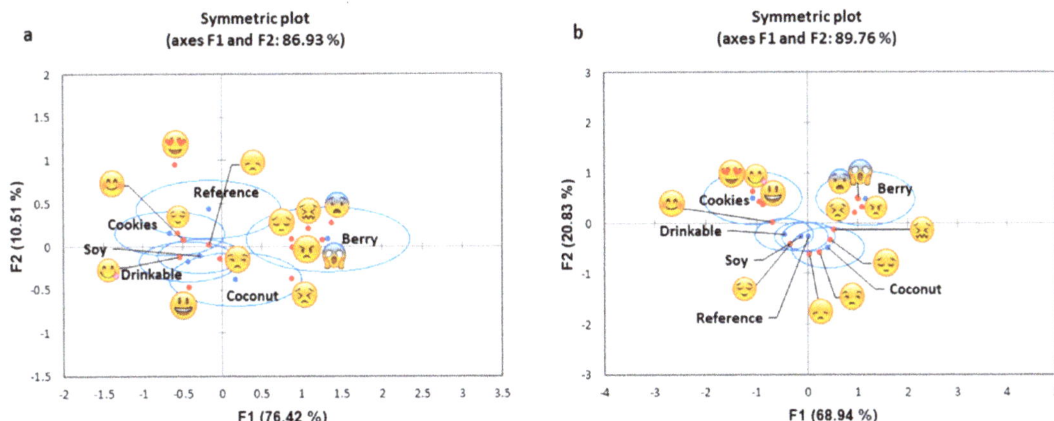

Figure 2. Correspondence analysis symmetric plots for emojis comparing for the two cultural groups: (**a**) Western consumers, (**b**) Asian consumers, with the confidence ellipses showing emojis related to each of the yogurt products.

3.4. Comparison of Facial Expression Recognition (FER) Method

Facial expressions of consumers in response to the different yogurt products were measured using the FaceReader[TM] software. Tables 4 and 5 show the mean values for the eight facial emotion terms, together with the valence and arousal estimates for Western and Asian consumers, respectively. No significant differences were observed among the FER attributes for Western consumers, whereas the emotions 'surprised' and 'disgusted' were significantly different ($p < 0.05$) for Asian consumers, where 'surprised' was rated highest for soy and lowest for berry, whereas 'disgusted' was rated highest for berry yogurt.

Table 4. Mean values for the biometric emotions for Western consumers.

Product Code	Neutral [NS]	Happy [NS]	Sad [NS]	Angry [NS]	Surprised [NS]	Scared [NS]	Disgusted [NS]	Contempt [NS]	Valence [NS]	Arousal [NS]
Reference	0.53 ± 0.21	0.02 ± 0.02	0.12 ± 0.18	0.12 ± 0.14	0.04 ± 0.05	0.02 ± 0.02	0.07 ± 0.11	0.01 ± 0.02	−0.23 ± 0.19	0.30 ± 0.11
Coconut	0.51 ± 0.20	0.03 ± 0.08	0.10 ± 0.14	0.14 ± 0.22	0.03 ± 0.04	0.02 ± 0.02	0.11 ± 0.13	0.01 ± 0.02	−0.24 ± 0.25	0.32 ± 0.12
Drinkable	0.51 ± 0.17	0.03 ± 0.10	0.16 ± 0.18	0.08 ± 0.10	0.02 ± 0.03	0.03 ± 0.04	0.11 ± 0.16	0.02 ± 0.03	−0.22 ± 0.23	0.27 ± 0.12
Soy	0.54 ± 0.21	0.03 ± 0.05	0.15 ± 0.19	0.11 ± 0.13	0.03 ± 0.04	0.02 ± 0.02	0.10 ± 0.13	0.01 ± 0.01	−0.25 ± 0.20	0.28 ± 0.11
Cookies	0.55 ± 0.18	0.01 ± 0.02	0.12 ± 0.15	0.13 ± 0.13	0.03 ± 0.05	0.02 ± 0.02	0.07 ± 0.08	0.01 ± 0.02	−0.22 ± 0.15	0.26 ± 0.12
Berry	0.47 ± 0.19	0.03 ± 0.04	0.11 ± 0.17	0.16 ± 0.18	0.03 ± 0.04	0.02 ± 0.02	0.12 ± 0.12	0.01 ± 0.02	−0.27 ± 0.20	0.33 ± 0.11
F-value	0.65	0.53	0.51	0.93	0.39	0.69	0.97	0.93	0.20	1.71

[NS] indicates no significant differences in that column ($p > 0.05$) by the Fisher's least square difference test. The data are presented as mean ± standard deviation.

Table 5. Mean values for the biometric emotions for Asian consumers.

Product Code	Neutral [NS]	Happy [NS]	Sad [NS]	Angry [NS]	Surprised	Scared [NS]	Disgusted	Contempt [NS]	Valence [NS]	Arousal [NS]
Reference	0.41 ± 0.14	0.05 ± 0.05	0.20 ± 0.17	0.10 ± 0.08	0.06 ± 0.10 [ab]	0.04 ± 0.04	0.09 ± 0.09 [b]	0.01 ± 0.02	−0.22 ± 0.17	0.30 ± 0.14
Coconut	0.41 ± 0.11	0.05 ± 0.05	0.19 ± 0.14	0.11 ± 0.12	0.06 ± 0.08 [ab]	0.05 ± 0.04	0.08 ± 0.07 [b]	0.01 ± 0.02	−0.22 ± 0.15	0.30 ± 0.12
Drinkable	0.40 ± 0.12	0.04 ± 0.05	0.23 ± 0.18	0.11 ± 0.14	0.08 ± 0.12 [ab]	0.03 ± 0.03	0.06 ± 0.08 [b]	0.01 ± 0.01	−0.26 ± 0.21	0.29 ± 0.10
Soy	0.38 ± 0.13	0.04 ± 0.06	0.25 ± 0.19	0.12 ± 0.13	0.09 ± 0.13 [a]	0.05 ± 0.06	0.05 ± 0.05 [b]	0.01 ± 0.01	−0.27 ± 0.19	0.29 ± 0.14
Cookies	0.40 ± 0.15	0.05 ± 0.06	0.23 ± 0.20	0.11 ± 0.12	0.05 ± 0.07 [ab]	0.05 ± 0.04	0.04 ± 0.05 [b]	0.01 ± 0.04	−0.23 ± 0.22	0.82 ± 2.84
Berry	0.36 ± 0.11	0.03 ± 0.04	0.20 ± 0.16	0.12 ± 0.13	0.03 ± 0.04 [b]	0.04 ± 0.05	0.15 ± 0.17 [a]	0.01 ± 0.01	−0.31 ± 0.15	0.33 ± 0.17
F-value	0.69	0.68	0.59	0.16	1.34	0.65	5.26	0.11	1.07	1.03

[a,b] Means with different superscripts in each column indicate significant differences ($p < 0.05$) by the Fisher's least square difference test. [NS] indicates no significant differences in that column ($p > 0.05$). Highest value indicated with 'a' in superscript. The data are presented as mean ± standard deviation.

The PCA biplot (Figure 3) shows that 86.35% of the variability in Western consumers' responses could be attributed to the two principal components (F1: 50.25%, F2: 36.11%) and 73.28% of the variability could similarly be attributed to Asian consumers (F1: 47.52%, F2: 25.76%), respectively. 'Happy' was positively related to 'disgusted' and was negatively related to 'surprised' and 'neutral' in the case of Western consumers. In contrast, 'happy' was positively related to 'neutral' and 'valence' in the case of Asian participants and was negatively related to 'angry'. The coconut and berry incited 'arousal' in the case of Western consumers and were linked to 'angry' emotion. The coconut was related to 'contempt' and berry was placed in between 'angry' and 'disgusted' for Asian consumers. The drinkable and cookies incited 'arousal' in the case of Asians and were linked to 'surprised' and 'scared' emotions.

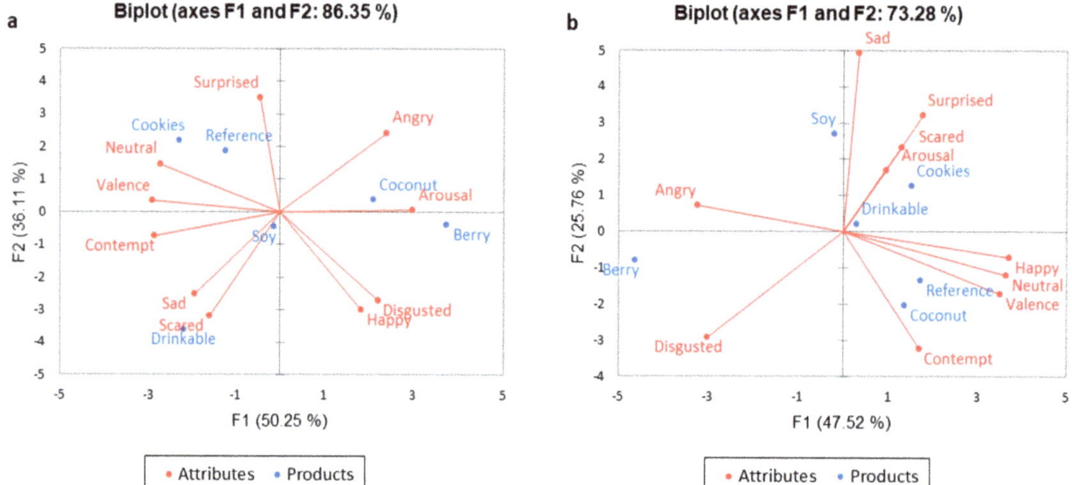

Figure 3. Biplots for facial expression recognition analysis comparing for the two cultural groups: (**a**) Western consumers, (**b**) Asian consumers.

3.5. Comparison and Relationship between Methods

The three methods—CATA emotions, CATA emojis and facial expression recognition—were examined using a linear mixed model, with overall liking as the response variable and the individual attributes as fixed effects. The percentage variance explained by each model was calculated by fitting the same explanatory variables in a linear regression model. The difference in means and standard error for each of the term in the models is shown in Table 6.

CATA emotions explained 66.0% of the variation, with 'cheerful', 'neutral', 'trusted', 'indifferent' and 'uplifting' being positively correlated with overall liking and showing an increase in the mean rating value of 2.29, 0.84, 1.16, 0.93 and 0.81, respectively. The terms 'artificial' and 'nasty' decreased the mean rating value by 0.87 and 2.31 units, respectively, and were negatively correlated with overall liking.

The CATA emojis accounted for 67.8% of the overall variance with the icons, with 'heart-shaped eyes 😍', 'relieved face 😌', 'smiling face 😊' and 'smiling face with mouth open 😃' positively related to overall liking, showing an increase in the mean rating value of 2.08, 1.28, 1.90 and 1.35, respectively. The other emojis, including 'angry face 😠', 'fearful 😨', 'pensive face 😔', 'persevering face 😣', 'confounded face 😖' and 'screaming in fear 😱', were negatively linked to overall liking, with a reduction in the mean rating value by 1.88, 1.69, 1.32, 1.60, 1.05 and 0.13 units, respectively. Only model 2 showed a significant effect ($p < 0.05$) of culture (Asian versus Western consumers). Western consumers had a decrease in the overall liking score of 0.36 in the model compared to Asian consumers. In other models (1 and 3), culture was not a significant ($p > 0.05$) effect.

Table 6. A linear mixed model for each of the methods, taking 'overall liking' as the response variable and representing percentage variance explained by each method.

Model	Fixed Factor	Difference between Means	Standard Error (SE)	p Value	F-Value	Random Factor	Overall Mean	Standard Error of Difference	Percentage Variance
Check-all-that-apply (CATA) emotions	Artificial	−0.87	0.18	<0.001	24.16	Yogurt Samples	4.91	0.20	66
	Cheerful	2.29	0.21	<0.001	116.49				
	Nasty	−2.31	0.22	<0.001	106.11				
	Neutral	0.84	0.21	<0.001	15.86				
	Trusted	1.16	0.25	<0.001	21.77				
	Uplifting	0.81	0.23	<0.001	11.17				
	Indifferent	0.93	0.28	0.001	10.67				
Check-all-that-apply (CATA) emojis	😠	−1.88	0.38	<0.001	24.59	None	4.91	0.14	67.8
	😨	1.69	0.27	<0.001	38.15				
	😍	2.08	0.38	<0.001	30.58				
	😔	−1.32	0.27	<0.001	23.86				
	😣	−1.60	0.25	<0.001	39.28				
	😌	1.28	0.19	<0.001	43.97				
	😊	1.90	0.19	<0.001	101.36				
	😃	1.35	0.25	<0.001	29.30				
	😖	−1.05	0.26	<0.001	16.28				
	😱	−0.13	0.37	<0.001	12.49				
FER	Culture	−0.36	0.15	0.019	5.52	Yogurt Samples	5.52	0.29	8.8
	Disgusted	−4.73	1.15	<0.001	11.11				
	Surprised	5.73	1.72	<0.001	16.94				

The facial emotion recognition explained only 8.8% of the variation, with the term 'surprised' linked positively to overall liking, with an increase of 5.73 units in the mean value, and the term 'disgusted' was negatively linked to overall liking, decreasing the mean rating value by 4.73 units.

The multifactorial analysis (MFA) plot combined all three data sets, estimating how closely each factor was linked to other factors and the yogurt products. The MFA plot for all participants (Figure 4) explained 90.43% of the total variability (F1: 80.66%, F2: 9.76%). Overall liking was positively linked to 'cheerful', 'uplifting' and 'trusted' in case of the emotion terms, 'smiling face 😊', 'relieved face 😌', 'smiling face with mouth open 😃' and 'heart-shaped eyes 😍' in case of the emojis, and 'surprised' in case of the biometrics. The overall liking was negatively linked to 'artificial', 'nasty', 'indifferent' and 'neutral' emotion terms, 'fearful 😨', 'persevering face 😣', 'angry face 😠', 'pensive face 😔', 'screaming in fear 😱' and 'confounded face 😖' emojis, and 'disgusted' for the biometrics measurements. The soy, drinkable, reference and cookies were all positively linked to the overall liking, and the berry and coconut were linked to negative terms.

The MFA plots comparing the responses of Western and Asian participants (Supplementary Figures S1 and S2) showed that the most liked yogurt (cookies) and the least liked yogurts (berry and coconut) were similarly placed across both the cultures. Apart from this, soy and drinkable showed minor differences (although positive) across the two groups.

However, the reference yogurt was related to more positive terms by Western participants and associated with slight dislike by Asians.

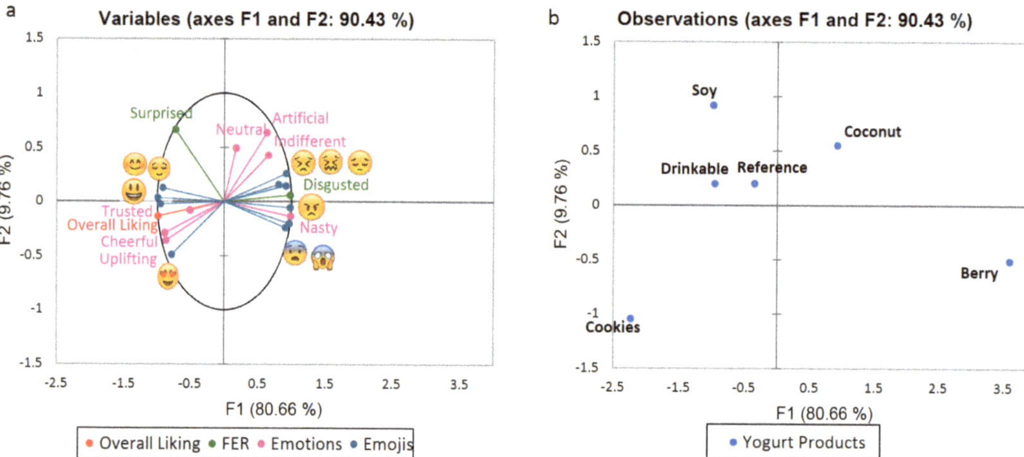

Figure 4. Multifactor analysis for a combination of check-all-that-apply (CATA) emotions, check-all-that-apply (CATA) emojis and facial expression recognition (FER) with overall liking ((**a**)-variables, (**b**)-yogurt products).

3.6. Price Perception and Purchase Intent of Yogurts

Consumers perceived the cookies yogurt to be "somewhat" to "very likely" to be purchased from a supermarket, with an average purchase intent score of 3.6 (on a scale of 5), although they considered its price to be the same as the regular yogurts that they generally consume (rated by 59.7% participants). Only the berry yogurt received the lowest score of 1.1, indicating that it was 'not at all likely' to be purchased. The reference, drinkable and soy yogurts were rated to be in between 'not so likely' and 'somewhat likely', although their perceived price was rated to be the 'same' by 41.9, 38.7 and 35.5% of participants, as compared to their regular yogurts. The data are shown in Supplementary Tables S4 and S5. The interaction for culture and yogurt products was significant for purchase intent ($p < 0.05$). Asian consumers perceived the price of cookies and drinkable yogurts to be higher. Western consumers perceived the price of the reference yogurt to be higher.

4. Discussion

4.1. Traditional Sensory Method

The advantage of using a 9-point hedonic scale is that it is simple to use by consumers and is effective in predicting the acceptance of a food product [2]. In the present study, it was successfully used to predict the liking of the six different yogurt formulations, suggesting that the cookies was the most liked and the berry was the least liked yogurt product. However, these scales could not predict any differences in the emotional responses of participants, or differences in liking across cultures, which is in contrast to some of our previous works, in which cultural differences influenced the affective responses towards different product types [29].

The hedonic scales have some drawbacks, including less freedom for the participants to express a full range of emotions [2]. Furthermore, these liking scores are limited in predicting market success. Hence, more elaborate methods, that measure consumer emotions (conscious or unconscious), need to be included for a more complete understanding of consumer food choices [30].

4.2. Self-Reported Responses (CATA Methods)

Apart from hedonic scale, self-reported methods of emotions or emojis has gained popularity for sensory segmentation of emotions in sensory experiments. This study shows that this approach allows differences in liking towards the same yogurt products to be better understood and is particularly useful where Western and Asian consumers use different terms to express liking towards the same yogurt type. For example, in this study, Asians selected 'common' to describe the reference yogurt, and the berry yogurt was associated more with the 'pensive face 😔' by Western consumers. A similar effect was found for the selection of the emotional terms in a study of chocolate tasting, as Asian and Western consumers expressed emotions relating to the consumption of chocolate differently [28]. Hence, CATA terms, when used in combination with a hedonic scale, can better explain consumer preferences and segmentation, without creating any bias [31].

Further, it has been found that emojis and emoticons can be an easy and effective way to understand food-related emotions, as consumers can relate to visual cues better than words [12]. Jaeger, Xia, Lee, Hunter, Beresford and Ares [7] found emojis (or emoticons) to be suitable for understanding the liking of a wide variety of food products by consumers. In another study by Jaeger et al. [32], emojis were successfully used to assess emotional liking towards the consumed food products (where a variety of products were served according to culture), and emojis had the potential to be used as an approach for direct measurement.

There are limitations of the self-reported methodology of emojis, as subtle differences between the samples cannot be easily detected, although this method does offer a simple and easy approach for understanding the sensory characteristics of food products [13]. Furthermore, it is important to define an emotion, so that it is easier to understand consumers' expectations from the tasted product and also to keep a balance of positive and negative emotion terms, in order to eliminate bias [33].

4.3. Facial Expression Recognition Responses (FER Method)

To overcome the bias associated with self-reporting sensory terms, an unconscious method was developed that involves the recording of the facial expressions of consumers, without informing them (intrinsically). In the yogurt study, the most liked product (cookies) was related to the 'surprised' facial expression emotion. The least liked product (berry) was related to the 'disgusted' facial emotion by both cultures. 'Happy' was not an effective discriminator for the products based on facial expressions, as it was related to 'disgusted' in the case of Western participants and to 'neutral' for Asian participants. Previously, in a study tasting orange juices, the 'neutral', 'disgusted' and 'angry' facial expressions explained liking towards the tasted samples, whereas the 'happy' emotion was not a useful discriminator during the implicit measurement of samples. It was also shown in this implicit study that participants displayed more negative emotions, and the least liked samples were easier to differentiate [34], consistent with observations made here.

The FaceReader™ emotions detected clear cultural differences here among Asian consumers, who showed a significant difference in means for 'surprised' and 'disgusted', whereas no differences were observed for Western consumers. A similar outcome was previously observed for chocolate samples, where no significant differences were observed by FaceReader™ emotions, although the consumer panel was not differentiated by culture [21].

The facial expression recognition method also has some limitations. The mean value for the term 'neutral' was the highest for both cultures. This could be attributed to the fact that participants performed tastings in an isolated booth, which was a socially distant and neutral environment. Hence, the consumers did not show much variation in positive emotions, but negative emotions were more pronounced. In a related study, it was seen that liking was not well correlated with emotional effects but a shift in liking for the yogurts tasted was well distinguished, showing the importance of emotions experienced [35]. Negative emotions were also better displayed for disliked samples in another study of different juices [36]. This suggests that it is easier to distinguish a disliked product using facial recognition. Alvarez-Pato et al. [37] also found that facial recognition could not be used as a stand-alone method for predicting consumer acceptance of food products and this method is better applied in combination with other techniques, which is consistent with the findings here.

4.4. Comparison and Relationship of Methods

To compare the conscious methods of sensory analysis with the unconscious method, the overall liking scores (from hedonic scales) were used as a reference method. This comparison was performed to determine the method that best represents the variance observed and to provide the most accurate consumer insights.

The conscious methods, CATA emotions and emojis, explained more variation in the consumer data—67.8% and 66.0%, respectively. The emojis model also explained the variation for the cultural differences ($p < 0.05$), which could not be explained by the other two models. The unconscious method (FER) explained quite a low level of variation within the data (8.8%). Only two terms were relevant in the model, which were 'surprised' and 'disgusted'. This outcome suggests that self-reported methods can be used alone with a hedonic scale to explain the variability in consumer response. In contrast, facial emotion recognition should be used in combination with other techniques for a better estimation of variance across consumer responses.

The MFA plot, shown in Figure 4, explains the relationship between the hedonic scales and conscious and unconscious methods of sensory analysis for the six yogurt types. The plot shows an association of positive terms with the overall liking scores and shows negative terms to be opposite to liking. In a beer tasting study [22], unconscious biometric responses were successfully integrated with conscious sensory responses for predicting consumer liking of beers. However, in another study by van Bommel et al. [38] where implicit facial expressions were compared with the explicit self-reported emotions, little overlap was found between methods, which were not directly comparable. A sensory study of organic herbal infusions [8] also found similarities between the conscious responses chosen by the consumers (EsSense Profile®) and unconscious responses from their facial expressions and successfully characterized emotional responses. Negative emotions were better identified by FaceReader™ and positive emotions were better displayed with an EsSense Profile®. However, there are differing responses in the literature, as another study, involving a comparison of self-reported and implicit responses for beer, found liking scores to have a higher discrimination compared to facial expression and also found self-reported emotional responses to display the highest discrimination compared to the other two methods [39]. Another study with juices also found that a higher correlation of negative emotions from facial expression analysis was observed with hedonic liking measurements [40].

The price perception in this study for the most liked yogurt (cookies) and least liked yogurt (berries) was directly related to the liking predicted by all three sensory methods. The plain plant soy yogurt and the reference dairy yogurt were similarly liked by the participants, indicating that liking does not necessarily depend on the source of protein. However, the different food textures of the yogurts evoked different emotions in our study. Reports have shown that emotions are more linked to the intrinsic properties of a food product and can vary accordingly, even if the liking scores are the same. The relation between the emotions and liking is not straightforward [41], also confirmed in our study with yogurts. Further, familiarity or neophobia is an important parameter that plays an significant role in the representation of emotions across cultures [42]. In a study with Greek yogurts, different textures produced varying consumer responses and liking [43]. Hence, it is important to understand the emotions linked to the tasted yogurts, as in our case, to understand food choice. Studies have shown that different cultures understand the same set of emotions, expressing these differently [33]. Other structural and compositional factors may also affect liking scores, which were not explored in detail in this study.

5. Conclusions

The traditional sensory methods using a hedonic scale offer benefits in terms of their speed and ease of use but have shortcomings in quantifying the full range of emotional responses and reasons behind liking or disliking of yogurt products. Conscious and unconscious methods can also be combined with a hedonic scale for a better understanding of consumer perceptions. These advanced approaches were found to be useful in measuring the cultural differences in liking towards yogurt products, even though the hedonic scale liking scores were similar. The conscious approach using the CATA methodology explained a higher proportion of variance for yogurt-tasting data. In contrast, the novel facial expression recognition approach distinguished the acceptability of different yogurt samples—more specifically, disliked samples—even though this method explained a very low proportion of the variability in the consumer data. The effect of culture was shown by the CATA and FER methods, where each culture displayed a different set of emotions for the tasted yogurts. Although an effect of culture was not shown by the traditional hedonic scale method, this effect was seen on the price perception of the tasted yogurts. At least in the evaluation of yogurt formulations, it is therefore recommended that these different methods be combined for a more comprehensive explanation of consumer perceptions and expectations in a multi-cultural environment.

6. Limitations and Future Research

The limited sample size in the experiment may present certain constraints in interpreting the results. Therefore, future research is recommended by increasing the sample size of participants. Additionally, the dependence of gender can be studied through all three methods tested in the experiment once the sample size is increased. Furthermore, the sensitivity of tasting can be checked for a second and third tasting trial using the FER method (that is, having multiple tastings during the sensory session) to understand how the result could vary with more than one trial, for the same participant. The FER can also be tested in different locations for a more ecologically valid experience, rather than a closed isolated setting, which might help in generating fewer facial expressions. Product parameters, such as particle size and emulsifiers, are also important factors contributing to the liking of yogurts, and these parameters can be measured and linked to the emotions to improve our understanding of liking in further studies.

Supplementary Materials: The following are available online at https://www.mdpi.com/article/10.3390/foods10061237/s1. Table S1: List of check-all-that-apply (CATA) emotion terms given to the consumers. Table S2: List of check-all-that-apply (CATA) emojis given to the consumers. Table S3: List of emotions (unconscious responses) represented by FaceReader[TM]. Table S4: Purchase intent of yogurts as rated by participants. Table S5: Price perception of yogurts as rated by participants in reference to regular yogurts (all values are in percentages).

Author Contributions: Conceptualization, M.G., D.D.T., F.R.D. and J.J.C.; methodology M.G. and D.D.T.; formal analysis, M.G. and D.D.T.; investigation, M.G., D.D.T. and G.H.; resources, J.J.C. and F.R.D.; data curation, M.G., D.D.T. and J.J.C.; writing—original draft preparation, M.G.; writing—review and editing, M.G., D.D.T., J.J.C., G.H., L.O., S.L.G. and F.R.D.; visualization, M.G. and D.D.T.; supervision, D.D.T., J.J.C., S.L.G. and F.R.D.; project administration, J.J.C. and F.R.D.; funding acquisition, J.J.C. and F.R.D. All authors have read and agreed to the published version of the manuscript.

Funding: This research was funded by a Melbourne Research Studentship from the University of Melbourne and the "Future Food Hallmark Research Initiative project" of the University of Melbourne.

Institutional Review Board Statement: The study was approved by the Human Ethics Advisory Group (HEAG) of the Faculty of Veterinary and Agriculture Sciences (FVAS) at the University of Melbourne (Ethics ID 1853507.2) on 22 August 2019.

Informed Consent Statement: Informed consent was obtained from all subjects involved in the study.

Acknowledgments: The authors would like to acknowledge the assistance of colleagues, including, Xinyu Miao, Behannis Mena, Melindee Hastie, Maryam Paykary, Zhenzhao Li and Claudia Gonzalez for helping in the sensory sessions. A special thanks to Hollis Ashman for helping with planning of the sensory study.

Conflicts of Interest: The authors declare no conflict of interest.

References

1. Iannario, M.; Manisera, M.; Piccolo, D.; Zuccolotto, P. Sensory analysis in the food industry as a tool for marketing decisions. *Adv. Data Anal. Classif.* **2012**, *6*, 303–321. [CrossRef]
2. Lim, J. Hedonic scaling: A review of methods and theory. *Food Qual. Prefer.* **2011**, *22*, 733–747. [CrossRef]
3. Prescott, J.; Bell, G. Cross-cultural determinants of food acceptability: Recent research on sensory perceptions and preferences. *Trends Food Sci. Technol.* **1995**, *6*, 201–205. [CrossRef]
4. He, W.; Boesveldt, S.; de Graaf, C.; de Wijk, R.A. The relation between continuous and discrete emotional responses to food odors with facial expressions and non-verbal reports. *Food Qual. Prefer.* **2016**, *48*, 130–137. [CrossRef]
5. Ng, M.; Chaya, C.; Hort, J. Beyond liking: Comparing the measurement of emotional response using EsSense Profile and consumer defined check-all-that-apply methodologies. *Food Qual. Prefer.* **2013**, *28*, 193–205. [CrossRef]
6. Meiselman, H.L. The future in sensory/consumer research: Evolving to a better science. *Food Qual. Prefer.* **2013**, *27*, 208–214. [CrossRef]
7. Jaeger, S.R.; Xia, Y.; Lee, P.; Hunter, D.C.; Beresford, M.K.; Ares, G. Emoji questionnaires can be used with a range of population segments: Findings relating to age, gender and frequency of emoji/emoticon use. *Food Qual. Prefer.* **2018**, *68*, 397–410. [CrossRef]
8. Rocha, C.; Lima, R.; Moura, A.; Costa, T.; Cunha, L. Implicit evaluation of the emotional response to premium organic herbal infusions through a temporal dominance approach: Development of the temporal dominance of facial emotions (TDFE). *Food Qual. Prefer.* **2019**, *76*, 71–80. [CrossRef]
9. Ares, G.; Barreiro, C.; Deliza, R.; Giménez, A.; Gámbaro, A. Application of a check-all-that-apply question to the development of chocolate milk desserts. *J. Sens. Stud.* **2010**, *25*, 67–86. [CrossRef]
10. Giacalone, D.; Bredie, W.; Frøst, M.B. "All-In-One Test" (AI1): A rapid and easily applicable approach to consumer product testing. *Food Qual. Prefer.* **2013**, *27*, 108–119. [CrossRef]
11. Son, J.-S.; Do, V.B.; Kim, K.-O.; Cho, M.S.; Suwonsichon, T.; Valentin, D. Understanding the effect of culture on food representations using word associations: The case of "rice" and "good rice". *Food Qual. Prefer.* **2014**, *31*, 38–48. [CrossRef]
12. Vidal, L.; Ares, G.; Jaeger, S. Use of emoticon and emoji in tweets for food-related emotional expression. *Food Qual. Prefer.* **2016**, *49*, 119–128. [CrossRef]
13. Varela, P.; Ares, G. Sensory profiling, the blurred line between sensory and consumer science. A review of novel methods for product characterization. *Food Res. Int.* **2012**, *48*, 893–908. [CrossRef]
14. De Wijk, R.A.; He, W.; Mensink, M.G.J.; Verhoeven, R.H.G.; De Graaf, C. ANS Responses and Facial Expressions Differentiate between the Taste of Commercial Breakfast Drinks. *PLoS ONE* **2014**, *9*, e93823. [CrossRef]
15. Masupha, L.; Zuva, T.; Ngwira, S.; Esan, O. Face recognition techniques, their advantages, disadvantages and performance evaluation. In Proceedings of the 2015 International Conference on Computing, Communication and Security (ICCCS), Pamplemousses, Mauritius, 4–5 December 2015; pp. 1–5.
16. Bartkiene, E.; Steibliene, V.; Adomaitiene, V.; Juodeikiene, G.; Cernauskas, D.; Lele, V.; Klupsaite, D.; Zadeike, D.; Jarutiene, L.; Guiné, R.P.F. Factors Affecting Consumer Food Preferences: Food Taste and Depression-Based Evoked Emotional Expressions with the Use of Face Reading Technology. *BioMed Res. Int.* **2019**, *2019*, 1–10. [CrossRef]
17. Terzis, V.; Moridis, C.N.; Economides, A.A. Measuring instant emotions based on facial expressions during computer-based assessment. *Pers. Ubiquitous Comput.* **2011**, *17*, 43–52. [CrossRef]
18. Jain, A.; Ross, A.; Prabhakar, S. An Introduction to Biometric Recognition. *IEEE Trans. Circuits Syst. Video Technol.* **2004**, *14*, 4–20. [CrossRef]
19. Liu, J.; Bech, A.C.; Waehrens, S.S.; Bredie, W.L. Perception and liking of yogurts with different degrees of granularity in relation to ethnicity, preferred oral processing and lingual tactile acuity. *Food Qual. Prefer.* **2021**, *90*, 104158. [CrossRef]
20. Viejo, C.G.; Zhang, H.; Khamly, A.; Xing, Y.; Fuentes, S. Coffee Label Assessment Using Sensory and Biometric Analysis of Self-Isolating Panelists through Videoconference. *Beverages* **2021**, *7*, 5. [CrossRef]
21. Gunaratne, T.M.; Fuentes, S.; Gunaratne, N.M.; Torrico, D.D.; Viejo, C.G.; Dunshea, F.R. Physiological Responses to Basic Tastes for Sensory Evaluation of Chocolate Using Biometric Techniques. *Foods* **2019**, *8*, 243. [CrossRef] [PubMed]
22. Viejo, C.G.; Fuentes, S.; Howell, K.; Torrico, D.D.; Dunshea, F.R. Integration of non-invasive biometrics with sensory analysis techniques to assess acceptability of beer by consumers. *Physiol. Behav.* **2019**, *200*, 139–147. [CrossRef]
23. Leitch, K.; Duncan, S.; O'Keefe, S.; Rudd, R.; Gallagher, D. Characterizing consumer emotional response to sweeteners using an emotion terminology questionnaire and facial expression analysis. *Food Res. Int.* **2015**, *76*, 283–292. [CrossRef]
24. McCarthy, K.; Parker, M.; Ameerally, A.; Drake, S.; Drake, M. Drivers of choice for fluid milk versus plant-based alternatives: What are consumer perceptions of fluid milk? *J. Dairy Sci.* **2017**, *100*, 6125–6138. [CrossRef]
25. Jeske, S.; Zannini, E.; Arendt, E.K. Past, present and future: The strength of plant-based dairy substitutes based on gluten-free raw materials. *Food Res. Int.* **2018**, *110*, 42–51. [CrossRef]

26. Fuentes, S.; Viejo, C.G.; Torrico, D.D.; Dunshea, F.R. Development of a Biosensory Computer Application to Assess Physiological and Emotional Responses from Sensory Panelists. *Sensors* **2018**, *18*, 2958. [CrossRef]
27. Gelici-Zeko, M.; Lutters, D.; Klooster, R.T.; Weijzen, P.L.G. Studying the Influence of Packaging Design on Consumer Perceptions (of Dairy Products) Using Categorizing and Perceptual Mapping. *Packag. Technol. Sci.* **2012**, *26*, 215–228. [CrossRef]
28. Gunaratne, T.M.; Viejo, C.G.; Fuentes, S.; Torrico, D.D.; Gunaratne, N.M.; Ashman, H.; Dunshea, F.R. Development of emotion lexicons to describe chocolate using the Check-All-That-Apply (CATA) methodology across Asian and Western groups. *Food Res. Int.* **2019**, *115*, 526–534. [CrossRef] [PubMed]
29. Torrico, D.D.; Fuentes, S.; Viejo, C.G.; Ashman, H.; Dunshea, F.R. Cross-cultural effects of food product familiarity on sensory acceptability and non-invasive physiological responses of consumers. *Food Res. Int.* **2019**, *115*, 439–450. [CrossRef]
30. Gutjar, S.; De Graaf, C.; Kooijman, V.; De Wijk, R.A.; Nys, A.; Ter Horst, G.J.; Jager, G. The role of emotions in food choice and liking. *Food Res. Int.* **2015**, *76*, 216–223. [CrossRef]
31. Ares, G.; Jaeger, S. Examination of sensory product characterization bias when check-all-that-apply (CATA) questions are used concurrently with hedonic assessments. *Food Qual. Prefer.* **2015**, *40*, 199–208. [CrossRef]
32. Jaeger, S.; Vidal, L.; Kam, K.; Ares, G. Can emoji be used as a direct method to measure emotional associations to food names? Preliminary investigations with consumers in USA and China. *Food Qual. Prefer.* **2017**, *56*, 38–48. [CrossRef]
33. Meiselman, H.L. A review of the current state of emotion research in product development. *Food Res. Int.* **2015**, *76*, 192–199. [CrossRef]
34. Danner, L.; Sidorkina, L.; Joechl, M.; Duerrschmid, K. Make a face! Implicit and explicit measurement of facial expressions elicited by orange juices using face reading technology. *Food Qual. Prefer.* **2014**, *32*, 167–172. [CrossRef]
35. Mojet, J.; Dürrschmid, K.; Danner, L.; Jöchl, M.; Heiniö, R.-L.; Holthuysen, N.; Köster, E. Are implicit emotion measurements evoked by food unrelated to liking? *Food Res. Int.* **2015**, *76*, 224–232. [CrossRef]
36. Danner, L.; Haindl, S.; Joechl, M.; Duerrschmid, K. Facial expressions and autonomous nervous system responses elicited by tasting different juices. *Food Res. Int.* **2014**, *64*, 81–90. [CrossRef] [PubMed]
37. Álvarez-Pato, V.M.; Sánchez, C.N.; Domínguez-Soberanes, J.; Méndoza-Pérez, D.E.; Velázquez, R. A Multisensor Data Fusion Approach for Predicting Consumer Acceptance of Food Products. *Foods* **2020**, *9*, 774. [CrossRef]
38. van Bommel, R.; Stieger, M.; Visalli, M.; de Wijk, R.; Jager, G. Does the face show what the mind tells? A comparison between dynamic emotions obtained from facial expressions and Temporal Dominance of Emotions (TDE). *Food Qual. Prefer.* **2020**, *85*, 103976. [CrossRef]
39. Beyts, C.; Chaya, C.; Dehrmann, F.; James, S.; Smart, K.; Hort, J. A comparison of self-reported emotional and implicit responses to aromas in beer. *Food Qual. Prefer.* **2017**, *59*, 68–80. [CrossRef]
40. Zhi, R.; Wan, J.; Zhang, D.; Li, W. Correlation between hedonic liking and facial expression measurement using dynamic affective response representation. *Food Res. Int.* **2018**, *108*, 237–245. [CrossRef] [PubMed]
41. King, S.C.; Meiselman, H.L. Development of a method to measure consumer emotions associated with foods. *Food Qual. Prefer.* **2010**, *21*, 168–177. [CrossRef]
42. Barrena, R.; Sánchez, M. Neophobia, personal consumer values and novel food acceptance. *Food Qual. Prefer.* **2013**, *27*, 72–84. [CrossRef]
43. Desai, N.; Shepard, L.; Drake, M. Sensory properties and drivers of liking for Greek yogurts. *J. Dairy Sci.* **2013**, *96*, 7454–7466. [CrossRef] [PubMed]

 *foods*

Article

Non-Invasive Biometrics and Machine Learning Modeling to Obtain Sensory and Emotional Responses from Panelists during Entomophagy

Sigfredo Fuentes *, Yin Y. Wong and Claudia Gonzalez Viejo

Digital Agriculture, Food and Wine Sciences Group. School of Agriculture and Food,
Faculty of Veterinary and Agricultural Sciences, University of Melbourne, Parkville, VIC 3010, Australia;
awongyinying@gmail.com (Y.Y.W.); cgonzalez2@unimelb.edu.au (C.G.V.)
* Correspondence: sfuentes@unimelb.edu.au; Tel.: +61-03-9035-9670

Received: 22 June 2020; Accepted: 7 July 2020; Published: 9 July 2020

Abstract: Insect-based food products offer a more sustainable and environmentally friendly source of protein compared to plant and animal proteins. Entomophagy is less familiar for Non-Asian cultural backgrounds and is associated with emotions such as disgust and anger, which is the basis of neophobia towards these products. Tradicional sensory evaluation may offer some insights about the liking, visual, aroma, and tasting appreciation, and purchase intention of insect-based food products. However, more robust methods are required to assess these complex interactions with the emotional and subconscious responses related to cultural background. This study focused on the sensory and biometric responses of consumers towards insect-based food snacks and machine learning modeling. Results showed higher liking and emotional responses for those samples containing insects as ingredients (not visible) and with no insects. A lower liking and negative emotional responses were related to samples showing the insects. Artificial neural network models to assess liking based on biometric responses showed high accuracy for different cultures (>92%). A general model for all cultures with an 89% accuracy was also achieved.

Keywords: entomophagy; neophobia; alternative protein source; emotions; emojis

---

1. Introduction

Increasing public concerns related to the environmental impacts and health-related issues associated with foods from animal sources have influenced the increased research interest in the use and acceptability of alternative protein sources. Insects have been shown as a feasible, sustainable protein source in recent studies [1–5]. The latter is because, for animals, the production of 1 kg of animal weight will typically require around 2.5 kg of feed for chicken, 5 kg for pork, and 10 kg for beef. On the contrary, for insects, to produce the same weight of crickets (*Gryllidae*), around 1.7 kg of feed is required. Furthermore, edible insect parts are around 80% compared to only 55% for chicken and 40% for cattle [6]. Most insects, when compared to plants, have higher protein content—for example, insects have, in general, 35% to 77% protein content per edible weight, compared to soybeans with approximately 35%. Insects are considered as a complete protein food due to the presence and amount levels of all, or most of, the essential amino acids required for adequate human health that cannot be found from plant sources, such as cereals and legumes [7,8]. From an environmental point of view, insects can produce between 88% (cockroaches) and 46% less CO_2 (crickets and beetles) compared to beef cattle per unit of weight gain. Furthermore, the production of other greenhouse gases with 300 and 84 times more potency compared to CO_2, such as N_2O and CH_4, respectively, are negligible compared to those produced by beef cattle or pigs [9].

Although the consumption of insects dates from ancient times and up to 100 million years ago from anthropological studies, and they are currently a common source of food in more than 100 countries [10,11], most cultures are still reluctant to include them as part of their daily diets [3]. In previous studies, food neophobia and disgust have been reported as the main reasons for consumers being hesitant to try insect-based foods, especially for Westerners [1]. Other studies have reported that curiosity, novelty, and interest in searching for healthier meat alternatives are the main drives for consumers to try insect-based foods [12,13]. It has also been found that consumers are more willing to try these foods when the insects are used as part of their ingredients and are not visible, compared to when they can see the whole insect or parts of them within the presented dish [1,14]. Therefore, the initial assessment of the visual attributes of dishes of insect-based foods is very important, as they create the first impression for consumers and determine their eagerness to taste the product or not. Furthermore, the presentation of food, beverages, and even the packaging of food products using imagery presented on digital screens renders statistically similar information when presenting the same product for taste or handling, as it creates the first impression for consumers when judging a product [15–17]. Hence, the visual renderings of insect-based food may help to break negative emotions related to first impressions.

Due to the influence of emotions in decision-making, especially for new food products and even food packaging for consumer acceptability [15,16,18–21], it is important to assess how insect-based food products make consumers feel, besides studying only their acceptability. Therefore, some studies have been conducted to evaluate emotional responses towards foods with insects [22], specifically focused on disgust [23,24]. More recently, the use of emojis has been implemented for this purpose, as consumers tend to identify emotions using proxy images related to expressed emotions as non-verbal cues, as these are more closely associated with their feelings [25–27]. Subconscious responses from consumers have also been evaluated for chocolate and beer using computer vision techniques and machine learning to assess their facial expressions [15,16,20,28,29]. These studies may produce more relevant information that may be missed using conventional self-reported sensory analysis, especially for insect-based food products.

Therefore, this study aimed to assess consumers' self-reported and subconscious emotional responses towards different insect-based food samples evaluated using emojis, video capturing, and analysis (biometrics) as a quick and reliable approach to understand emotional responses according to cultural backgrounds. To achieve this, a cross-cultural analysis was conducted to compare the responses of Chinese and Australian participants. A sensory session was conducted using the BioSensory application (app; University of Melbourne, Melbourne, Vic, Australia; [30]) to obtain biometrics and check all that apply (CATA), using emojis to assess emotions and a 15 cm non-structured scale to evaluate the acceptability of five different foods prepared with insects. Furthermore, machine learning (ML) based on artificial neural networks (ANN) models were developed to classify samples into low and high overall liking using the subconscious (biometrics) emotional responses as inputs.

2. Materials and Methods

2.1. Sensory Session Description and Video Analysis

A total of 88 participants (34 Asians and 54 non-Asians) were recruited from the pool of staff and students at the University of Melbourne (UoM), Australia. The Asian participants were from countries such as China, Malaysia, Vietnam, Philippines, India, and Indonesia, while the non-Asians were from Australia, New Zealand, Mexico, Colombia, Germany, Ukraine, and the United States of America. The ethics considerations were related to the minimal risks associated with personal information and image/video acquisition from participants. Ethics approval was granted by the Faculty of Veterinary and Agricultural Sciences (FVAS) from the UoM with ID: 1545786.2. Furthermore, proper data handling and storage have been followed for security reasons and will be stored for five years. The participants were asked to sign a consent form for them to allow being video-recorded and to specify any allergies to assess whether they may take the test. The only information that was provided to participants about the samples along with the consent form previous to the test was "Some samples may contain insects", due to the allergic reactions that these may cause. A power analysis was performed using SAS® Power and Sample Size version 14.1 software (SAS Institute Inc., Cary, NC, USA), showing that the number of participants per culture was sufficient to find significant differences among the samples and cultures (Power: $1 - \beta > 0.99$). The session was conducted in individual sensory booths located in the sensory laboratory belonging to the FVAS of the UoM (Melbourne, Australia) using the BioSensory app [30] to display the questionnaire and automatically record videos of the participants while assessing the insect-based food samples. The participants were asked to taste two control and three different insect-based food samples (Table 1), one at a time, and rate their liking for different descriptors as well as to indicate how they felt towards the different samples using a FaceScale (FS; Table 2) and check all that apply (CATA) tests for different emojis that best represented their emotions towards the sample (Table 3). The samples were assigned a 3-digit random number and served at room temperature (20 °C); the avocado was prepared just before serving, and drops of lime juice were added to avoid oxidation and darkening. The participants were provided with water and water crackers as palate cleansers between samples.

Table 1. Image and description of samples used in the sensory session.

Sample Image	Sample Description
	Tortilla chip with cornflour (Control)
	Toast with avocado (Control)
	Tortilla chip with corn and cricket flour
	Toast with avocado and crickets
	Roasted crickets

Table 2. Descriptors and scale used in the questionnaire to acquire self-reported responses. Questionnaires were uploaded in the BioSensory app, including sample numbers, descriptors, scales, and emoticons.

Descriptor	Scale	Anchors	Label
Appearance	15 cm non-structured	Dislike extremely-Neither like nor dislike-Like extremely	Appearance
Appearance	FaceScale (0–100)		FS App
Aroma	15 cm non-structured	Dislike extremely-Neither like nor dislike-Like extremely	Aroma
Texture	15 cm non-structured	Dislike extremely-Neither like nor dislike-Like extremely	Texture

Table 2. *Cont.*

Descriptor	Scale	Anchors	Label
Flavor	15 cm non-structured	Dislike extremely-Neither like nor dislike-Like extremely	Flavor
Overall liking	15 cm non-structured	Dislike extremely-Neither like nor dislike-Like extremely	OL
Tasting	FaceScale (0–100)	(face scale emojis)	FS Taste
Purchase intention	15 cm non-structured	Dislike extremely-Neither like nor dislike-Like extremely	PI

Table 3. Emojis used for the check all that apply questions for the sensory test using the BioSensory app.

Emoji	Meaning	Emoji	Meaning
(emoji)	Happy	(emoji)	Savoring
(emoji)	Surprised	(emoji)	Scared
(emoji)	Expressionless	(emoji)	Angry
(emoji)	Disappointed	(emoji)	Confused
(emoji)	Neutral	(emoji)	Joy
(emoji)	Unamused	(emoji)	Laughing

Videos were recorded along with the self-reported answers from participants using the BioSensory app and analyzed using a second app developed with the Affectiva software development kit (SDK; Affectiva, Boston, MA, USA). The latter app can analyze videos in batch to obtain the facial expressions from participants based on the micro- and macro-movements of the different features of the face, using the histogram of the oriented gradient for computer vision analysis. Additionally, it uses machine learning algorithms based on support vector machine to translate facial expressions into emotions and emojis; it also assesses head movements (Table 4).

Table 4. Facial expressions and emotion-related parameters obtained from the biometric video analysis (Affectiva app).

Parameter	Label	Parameter	Label	Parameter	Label
Joy	Joy	Winking face	(emoji)	Lip Corner Depressor	LCD
Disgust	Disgust	Rage	(emoji)	Lip Press	LPr
Sadness	Sadness	Smirk	(emoji)	Lip Suck	LS

Table 4. *Cont.*

Parameter	Label	Parameter	Label	Parameter	Label
Surprise	Surprise	**Disappointed**		**Mouth Open**	MO
Anger	Anger	**Scared**		**Smirk Facial Expression**	SmirkFE
Fear	Fear	**Stuck out tongue with winking eye**		**Eye Closure**	EC
Contempt	Contempt	**Laughing**		**Eye Widen**	EW
Valence	Valence	**Kissing**		**Cheek Raise**	CR
Smile	Smile	**Inner Brow Raise**	IBR	**Lid Tighten**	LT
Engagement	Engagement	**Brow Rise**	BR	**Dimpler**	Dimpler
Attention	Attention	**Brow Furrow**	BF	**Lip Stretch**	LSt
Smiley		**Nose Wrinkle**	NW	**Jaw Drop**	JD
Relaxed		**Upper Lip Rise**	ULR	**Pitch**	
Stuck out tongue		**Chin Raise**	CR	**Yaw**	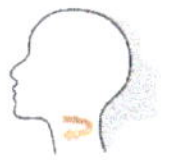
Flushed		**Lip Pucker**	LP	**Roll**	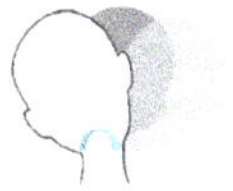

2.2. Statistical Analysis and Machine Learning Modeling

An ANOVA was conducted for the quantitative self-reported and biometric responses with a Fisher's least significant differences (LSD) *post hoc* test to assess significant differences ($\alpha = 0.05$) between samples nested within cultures (Asians and non-Asians) using XLSTAT ver. 2020.3.1 (Addinsoft, New York, NY, USA). Furthermore, multivariate data analysis was conducted using principal component analysis (PCA) to find the relationships and associations among samples and variables from the quantitative self-reported and biometric responses using Matlab® R2020a (Mathworks, Inc., Natick, MA, USA). On the other hand, to find relationships between the frequency responses from emojis using the CATA test and quantitative self-reported and biometric responses as well as the associations of samples with each response, a multiple factor analysis (MFA) was conducted using XLSTAT. Furthermore, a correlation matrix was developed using Matlab® R2020a to assess only the significant correlations ($p \leq 0.05$) between the quantitative self-reported and biometric responses.

The machine learning (ML) models based on artificial neural networks (ANN) pattern recognition were developed using a code written in Matlab® R2020a to evaluate 17 different training algorithms. Bayesian Regularization was selected as the best algorithm, resulting in models with higher accuracy and no signs of overfitting from performance tests. A total of 45 inputs from the biometrics emotion analysis (Table 4) were used to classify the samples into low and high overall liking. Model 1 was developed using data from Asian participants, while Model 2 was constructed using data from non-Asians. A further ML model (Model 3) was created as a general model, using the results from all the participants regardless of their cultural background. The data were divided randomly using 80% of the samples (samples x participants) for training and 20% for testing. The performance assessment was based on the mean squared error (MSE), and a neuron trimming test (3, 7, 10 neurons) was conducted to find the models with the best performance and no overfitting from performance tests. Figure 1 shows the diagram of the ANN models developed using a tan-sigmoid function in the hidden layer and Softmax neurons in the output layer.

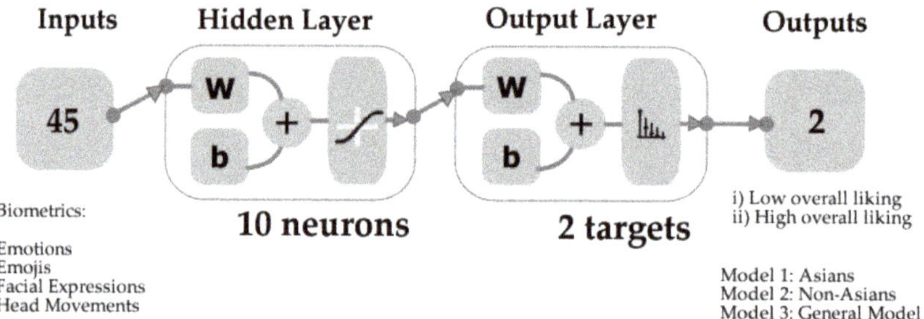

Figure 1. Diagram of the artificial neural network two-layer feed-forward models showing the number of inputs (Table 4), the outputs/targets, and the number of neurons.

3. Results

3.1. Results from the ANOVA of Self-Reported and Biometric Responses

Results from the ANOVA of the self-reported responses showed that there were significant differences between samples for the eight descriptors considered (Table S1). However, non-significant differences were found between the cultures for appearance, FS App, aroma, texture, and overall liking. For flavor, overall liking, and purchase intention, there were significant differences between cultures for the tortilla chips made with cricket flour and the avocado toast with crickets. Likewise, for FS Taste, there were significant differences between cultures for the avocado toast with crickets. The control samples of tortilla chips and avocado toast, along with the tortilla chips made with cricket flour, were the most liked in terms of appearance and texture. For Asians, besides the control samples, the tortilla chips made with cricket flour were the most liked in terms of texture and overall but were significantly different from the control. Conversely, for non-Asians, the tortilla chips made with cricket flour were amongst the highest in liking of all sensory descriptors as well as in the FaceScale rating, and non-significant differences were found with the control samples. Similarly, for non-Asians, the avocado toast with crickets was amongst the highest in flavor liking, with non-significant differences compared to the control samples.

The results of the ANOVA for the biometrics responses showed there were significant differences between the samples and/or cultures (Table S2). There were significant differences between the cultures for head roll  when assessing the avocado toast with crickets and the whole crickets, with more negative results for the Asians, which means that they moved the head to the right side when evaluating these samples. For joy, there were significant differences between cultures, with the avocado toast control sample having higher expression of this emotion in Asians. The Asians also behaved significantly differently to the non-Asians in terms of engagement when assessing the tortilla chips made with cricket flour, with the Asians being more engaged with the sample. The Asians expressed significantly more smiley faces when evaluating both the avocado toast control and the toast with crickets than the non-Asians. Likewise, the Asians expressed significantly more stuck out

tongue with winking eye 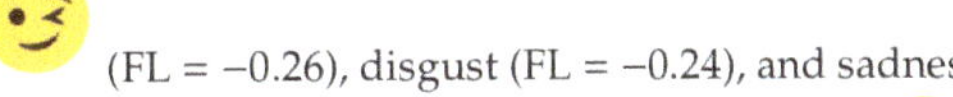 expressions when assessing the tortilla chips made with cricket flour.

3.2. Multivariate Data Analysis

3.2.1. Principal Components Analysis

Figure 2a shows the PCA from Asians, which explains 74.23% of the total data variability (PC1 = 44.18%; PC2: 30.05%). According to the factor loadings (FL), the principal component one (PC1) was mainly represented by roll (FL = 0.27), flavor (FL = 0.27), and texture (FL = 0.25) on the

positive side of the axis, and by winking face (FL = −0.26), disgust (FL = −0.24), and sadness

(FL = −0.24) on the negative side. On the other side, PC2 was mainly represented by kissing

(FL = 0.32), rage (FL = 0.30), and anger (FL = 0.30) on the positive side, and pitch (FL = −0.19) on the negative side of the axis. It can be observed that both control samples (avocado toast and tortilla chip) were associated with a higher liking of flavor, purchase intention, overall liking,

valence, smiley face , and roll 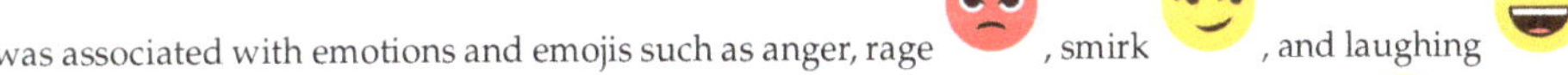, while the tortilla chip made with cricket flour

was associated with emotions and emojis such as anger, rage , smirk , and laughing .

On the other hand, the avocado toast with crickets was associated with winking face , disgust,

sadness, and flushed face , while the whole crickets were associated with disgust and pitch

 .

Figure 2b shows the PCA for non-Asians, which explained 70.43% of the total data variability (PC1 = 50.98%; PC2 = 19.45%). The PC1 was mainly represented on the positive side of the axis by

joy (FL = 0.25), smiley (FL = 0.25), engagement (FL = 0.24), and relaxed (FL = 0.24), and on the negative side by the self-reported responses (FL = 0.24) appearance, texture, flavor, overall

liking, and purchase intention. The PC2 was mainly represented by pitch (FL = 0.38) and

stuck out tongue with winking eye (FL = 0.21) on the positive side of the axis, and by surprise

(FL = −0.34), laughing (FL = −0.34), and disappointed (FL = −0.34) on the negative side.

The control sample of the tortilla chip was associated with yaw , and disappointed , while the control sample of avocado toast and tortilla chip with cricket flour were mainly linked with the liking of aroma and flavor, as well as disgust. On the other hand, the avocado toast with crickets was associated with pitch , stuck out tongue with winking eye , disgust, and smirk , while the whole crickets were linked with joy, valence, relaxed , anger, rage , engagement, and attention.

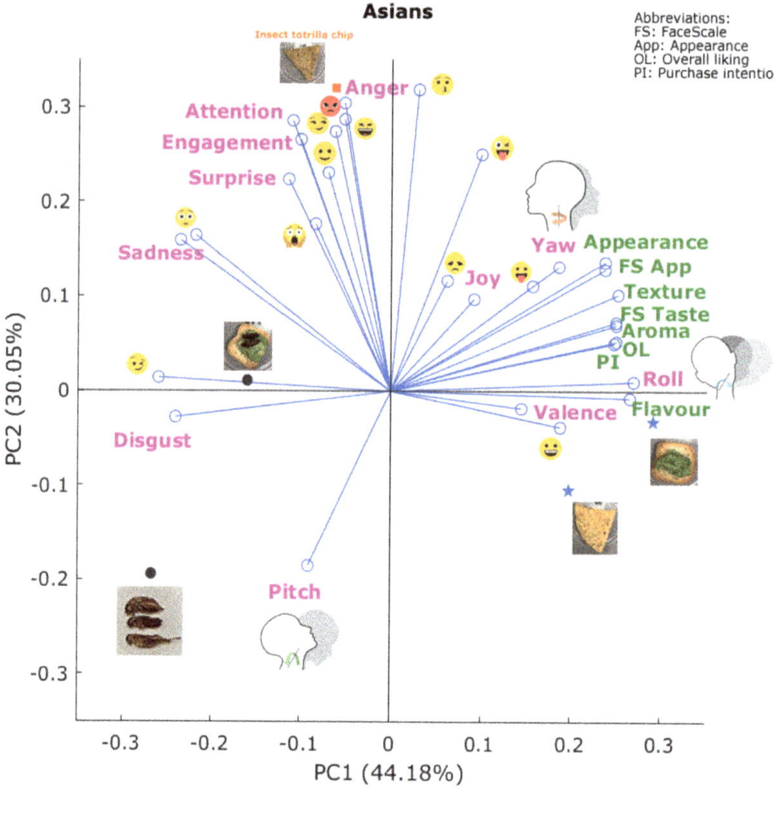

(a)

Figure 2. *Cont.*

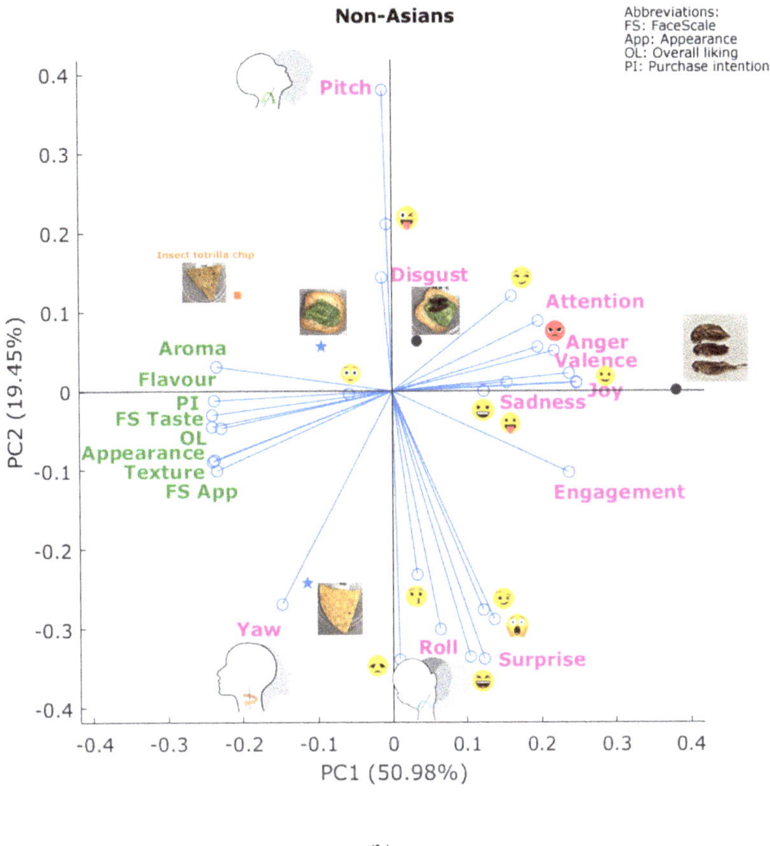

(b)

Figure 2. Principal components analysis for the biometric and quantitative self-reported responses for (**a**) Asians and (**b**) non-Asians. Abbreviations: PC1 and PC2: principal components one and two.

3.2.2. Correlation Analysis

Figure 3a shows the significant correlations ($p \leq 0.05$) between the quantitative self-reported and biometric emotional responses for Asians. It can be observed that roll was positively correlated with the liking of appearance (correlation coefficient: $r = 0.88$), aroma ($r = 0.92$), texture ($r = 0.93$), flavor ($r = 0.98$), overall liking ($r = 0.92$), FS Taste ($r = 0.91$), and purchase intention ($r = 0.91$). The liking of aroma was negatively correlated with winking face ($r = -0.94$), and disgust ($r = -0.95$), while the liking of flavor had a negative correlation with winking face ($r = 0.93$), and the latter with FS Taste ($r = -0.88$). On the other hand, Figure 3b shows the significant correlations ($p \leq 0.05$) found for non-Asians. It can be observed that joy had a negative correlation with the liking of appearance ($r = -0.94$), FS App ($r = -0.95$), the liking of texture ($r = -0.97$), flavor ($r = -0.89$), overall liking ($r = -0.94$), FS Taste ($r = -0.89$), and purchase intention ($r = -0.96$). Valence had a negative correlation with appearance liking ($r = -0.97$), FS App ($r = -0.97$), texture liking ($r = -0.96$), overall liking ($r = -0.95$), and FS Taste ($r = -0.96$). Similarly, relaxed was negatively correlated with appearance liking

($r = -0.92$), FS App ($r = -0.94$), texture liking ($r = -0.96$), overall liking ($r = -0.90$), and purchase intention ($r = -0.93$). Smiley has a negative correlation with appearance liking ($r = -0.89$), FS App ($r = -0.88$), liking of aroma ($r = -0.89$), texture ($r = -0.91$), flavor ($r = 0.92$), overall liking ($r = -0.90$), and purchase intention ($r = -0.90$).

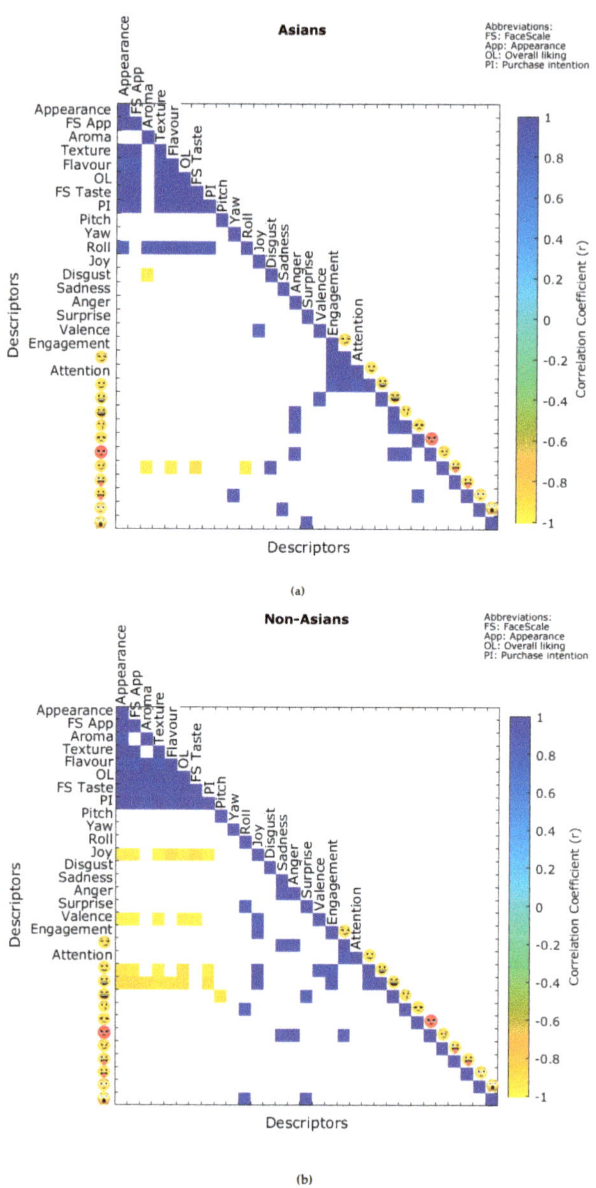

Figure 3. Matrices showing the significant ($p \leq 0.05$) correlations between the quantitative self-reported and biometric emotional responses for (**a**) Asians and (**b**) non-Asians. Color bar: blue side depicts the positive correlations, while the yellow side represents the negative correlations; likewise, darker blue and yellow denote higher correlations.

3.2.3. Multiple Factor Analysis

Figure 4a shows the MFA of Asian consumers, which explained a total of 76.61% (Factor 1: F1 = 53.69%; Factor 2: F2 = 22.92%). The F1 was mainly represented by surprised  (FL = 0.93), and laughing  (FL = 0.91) on the positive side of the axis, and by savoring 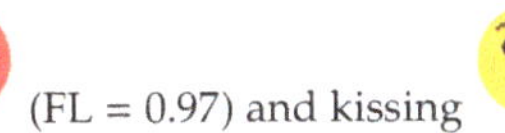 (FL = −1.00) and appearance liking (FL = −0.99) on the negative side. On the other hand, F2 was mainly represented on the positive side of the axis by rage  (FL = 0.97) and kissing (FL = 0.93), and on the negative side by unamused 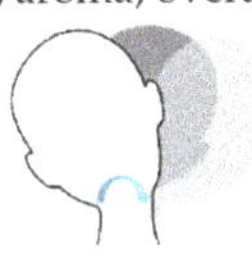 (FL = −0.49) and pitch (FL = −0.42). It can be observed that the control samples (tortilla chip and avocado toast) were associated mainly with self-reported responses such as the liking of flavor, aroma, overall liking, and purchase intention, as well as some subconscious responses such as roll , smiley, and valence. The tortilla chip made with cricket flour was associated with emojis measured with biometrics such as kissing , laughing 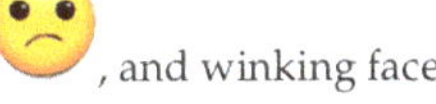, rage , and stuck out tongue with winking eye , and with neutral from the CATA test. On the other hand, the avocado toast with crickets was associated with emojis from the CATA test, such as disappointed, confused, and winking face 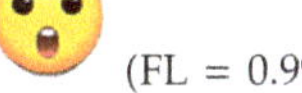 from the biometrics. The whole crickets were linked to emojis from the CATA test, such as laughing, surprised, expressionless, and pitch.

Figure 4b shows the MFA of non-Asian consumers, which explained a total of 76.59% (F1 = 58.26%; F2 = 18.33%). Based on the FL, the F1 was mainly represented by surprised 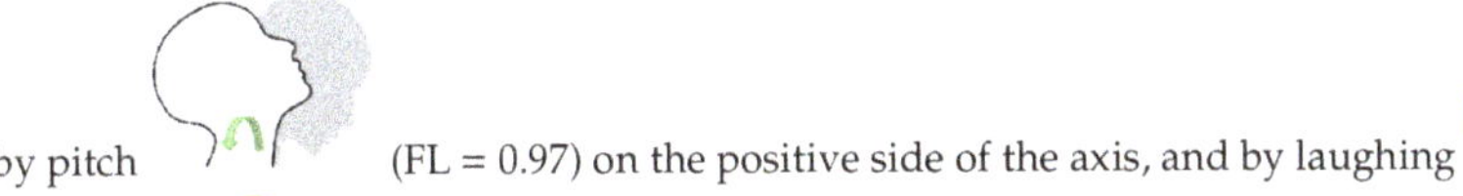 (FL = 0.99), disappointed  (FL = 0.99), and confused (FL = 0.99) on the positive side of the axis, and by texture liking (FL = −1.00), happy (FL = −0.99), liking of appearance (FL = −0.99), FS App (FL = −0.99), and purchase intention (FL = −0.99) on the negative side. On the other hand, F2 was represented by pitch (FL = 0.97) on the positive side of the axis, and by laughing (FL = −0.95) and scared (FL = −0.84) on the negative side. The control sample of the tortilla chip was mainly

Foods **2020**, 9, 903

associated with head movements such as yaw and roll and disappointed . The control sample of the avocado toast and the tortilla chip made with cricket flour were associated with self-reported responses such as the liking of aroma, flavor, overall liking, and purchase intention as well as flushed . The avocado toast with crickets was mainly associated with emojis from the CATA test, such as laughing , disappointed , unamused , confused , and neutral , and with biometric responses such as valence, stuck out tongue , and smirk . On the other hand, the whole crickets were linked to the biometric responses such as engagement, rage , winking face , and relaxed , and with emojis from the CATA test such as surprised and expressionless .

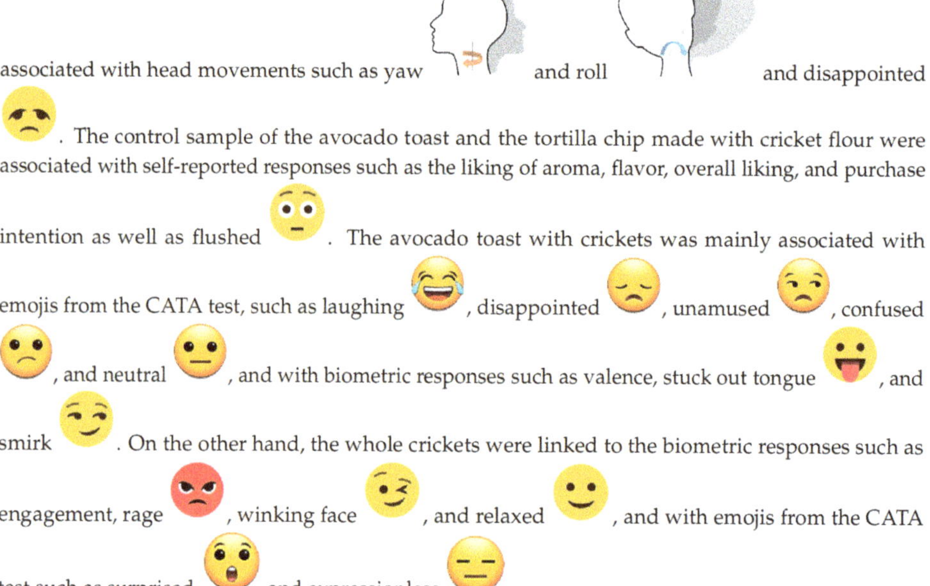

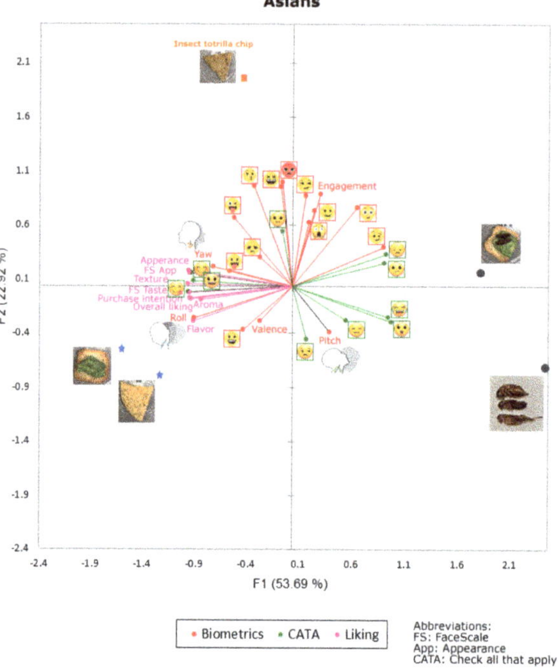

(a)

Figure 4. *Cont.*

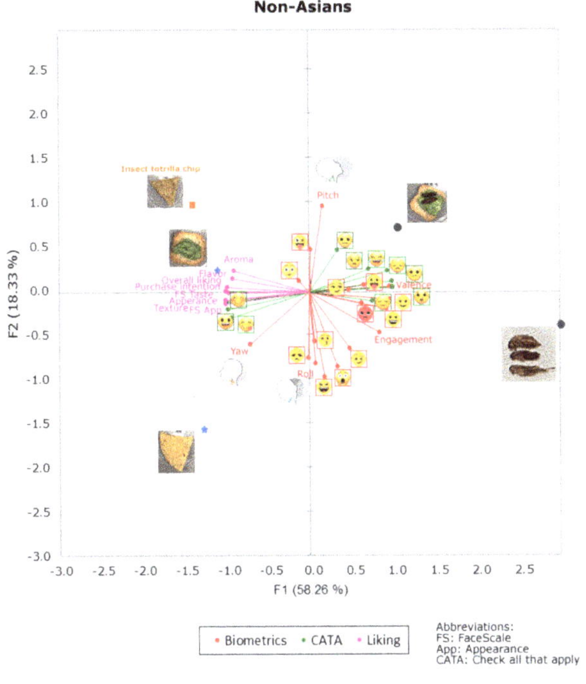

(b)

Figure 4. Multiple factor analysis for the biometric, frequencies (check all that apply: CATA) and quantitative self-reported (Liking) responses for (**a**) Asians and (**b**) non-Asians. Abbreviations: F1 and F2: factors one and two.

3.3. Machine Learning Modeling

In Table 5, it can be observed that Model 1, developed with the results from the Asian participants, had a 92% accuracy to classify the samples into high and low liking using only the emotional responses from biometrics as inputs. Model 2 for non-Asians had a higher overall accuracy (94%) compared to the Asians, and with higher accuracy also in the training stage (non-Asians: 76%; Asians: 71%). On the other hand, Model 3, which was developed as a general model with data from all the participants (Asians and non-Asians), presented an overall accuracy of 89%. The three models had a lower MSE value for the training stage compared to testing, which indicates that there was no overfitting. Figure 5 shows the receiver operating characteristics (ROC) curves, which depict the performance based on the specificity (false positive rate) and sensitivity (true positive rate) of each model and classification group.

Table 5. Statistical data from the three artificial neural network models. Performance shown is based on the mean squared error (MSE).

Stage	Samples × Participants	Accuracy	Error	Performance (MSE)
Model 1: Asians				
Training	134	97%	3%	0.03
Testing	34	71%	29%	0.24
Overall	168	92%	8%	-

Table 5. *Cont.*

Stage	Samples × Participants	Accuracy	Error	Performance (MSE)
Model 2: Non-Asians				
Training	216	99%	1%	0.01
Testing	54	76%	24%	0.20
Overall	270	94%	6%	-
Model 3: General				
Training	350	97%	3%	0.03
Testing	88	71%	29%	0.25
Overall	438	89%	11%	-

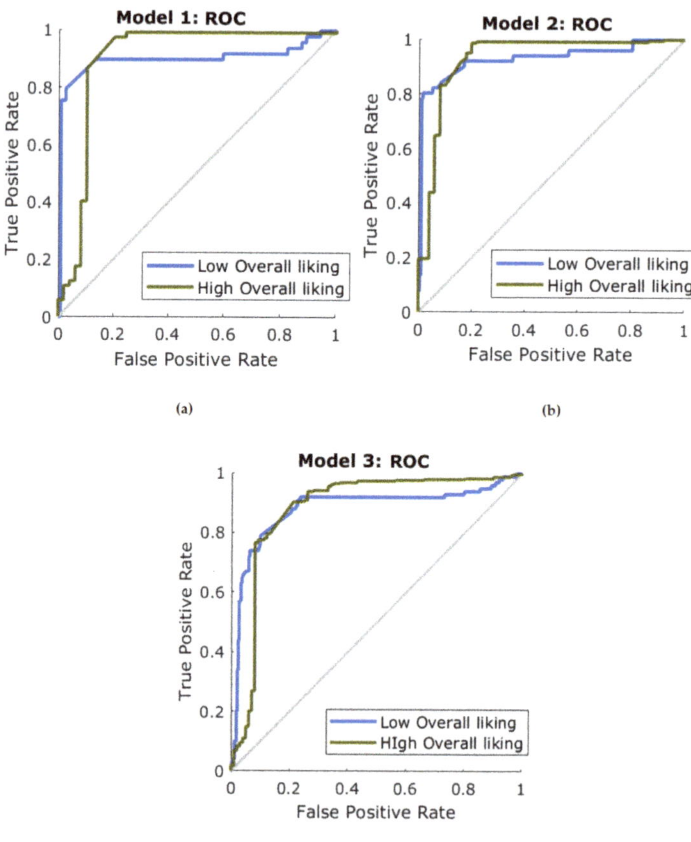

Figure 5. Receiver operating characteristics (ROC) curves for the three artificial neural network models for (**a**) Model 1: Asians, (**b**) Model 2: non-Asians, and (**c**) Model 3: general (Asians + non-Asians).

4. Discussion

In general, for this study, consumer perception had higher scores for self-reported information for the control samples, which did not contain insects, and the lowest scores for full insects, as expected. These scores were similar to the sample with insects as an ingredient, in which no parts of insects were visible, which is in accordance with previous literature [1,14,31]. However, mid-range liking, FS Taste, and purchase intention were obtained for visible insects mixed with more familiar products, such as toast and avocado mash (Table S1).

When combining self-reported information with the subconscious and emotional responses through the PCAs (Figure 2), it was found that high and similar variability of data was explained for both the Asian and non-Asian participants (>70%). For the Asians (Figure 2a), more negative emotional responses were associated with insect-based samples compared to the controls, with a clear separation in the vertical plane (PC1). For the non-Asians (Figure 2b), a similar separation was observed for the samples with visible insects, with a difference in the insect tortilla chip. These results may seem contrasting with the self-reported data; however, the subconscious responses are related to the first impression of the samples for visible, aroma, and taste responses from consumers, which are spontaneous and more complex when tasting food and beverage products of any kind [32–34]. Similar relationships were found using the correlation matrix analysis (Figure 3) and the further MFA combining the self-reported data, biometrics, and CATA.

A deeper understanding of consumer acceptability, liking, and intention to purchase new insect-based food products is extremely important, since around 95% of new food products may fail in the market without proper assessment [35,36]. Predictive modeling incorporating cultural backgrounds may offer more information and the possibility of automation for the decision-making process or product variation when developing new insect-based food products. Recent research has been based on automatic estimations of liking based on facial expression dynamics, especially for infants, since self-reported data may not be easily obtained [37,38]. However, research on automatic assessments based on biometrics is rarer [15,28,32].

By using ML modeling considering the separation of cultural background, it was possible to obtain high accuracy (≥92%) in the prediction of liking based on the biometrics from Asians (Table 5; Figure 5a) and non-Asians (Table 5; Figure 5b) using the Bayesian Regularization ANN algorithm. However, since ML using ANN is considered to be a robust method to detect patterns within data, the cultural distinction could be an internal feature of a general model, as shown in Table 5 and Figure 5c. This general model, with an accuracy of 89%, may be used to detect the liking levels of snacks containing insects as part of their ingredients or as whole insects. Further research is required to test and model the biometric responses from a wider variety of insect-based food or beverages.

A quicker analysis for consumer acceptability, liking, and intent to purchase can be achieved as a first approach by presenting images of dishes prepared with insects through the BioSensory app to obtain biometrics from panelists. It has been previously shown for packaging assessments that presenting images of the packaging, and the physical packaging samples resulted in non–significant statistical differences in their appreciation by panelists in a sensory trial [39]. Similarly, for beer tasting panelists looking at videos of beer pouring gave similar levels of liking and general appreciation compared to those that tasted the same beers [15,40]. The applicability for the assessment of images of insect-based dishes is supported by the data presented in this study, especially related to the statistically significant correlations between the appearance and overall liking for both cultures (Figure 3a,b) and the higher variability of the data explained by PC1 for both Asians and non-Asians (Figure 2a,b).

The advantages of using both the self-reported and biometric responses rely on the fact that they allow us to obtain the first and subconscious reaction that consumers have towards the product they are assessing as well as the way they may modify their responses after the thinking process. This aids in a deeper assessment of consumers' behavior and acceptability towards food and beverages to understand the target market and develop products according to their needs. The results from Asians and non-Asians were statistically analyzed separately due to the cultural differences that may be related to the expression of emotional responses. This was also the main reason for developing ML models for each culture and a general model considering both cultures. However, despite that Model 3 (general model) had a slightly lower accuracy, it was shown that ANN is able to accurately find patterns among data that may be related to differences in responses from different cultural backgrounds to predict liking regardless.

5. Conclusions

The introduction of insect-based food and beverage products in the market may be a viable alternative for sustainable and nutritional sources of protein. However, the neophobia associated with these products is based on the cultural background and lack of familiarity. More studies are required to break misconceptions and insect-based phobia, which requires the study of complex interactions between the response of consumers to insect-based food products due to visual first impressions, aroma, taste, emotional responses, and cultural background. The implementation of biometrics and machine learning modeling could provide deeper insights regarding these complex interactions and the liking/purchase intention of insect-based food products. Computer vision and machine learning should be further explored and researched as the basis for the development of an artificial intelligence approach to assess new food products and potential success in the market.

Supplementary Materials: The following are available online at http://www.mdpi.com/2304-8158/9/7/903/s1: Table S1: Means and standard error of the self-reported responses for each sample per culture. Different letters denote significant differences based on ANOVA and Fisher's least significant difference post hoc test ($\alpha = 0.05$). Table S2: Means and standard error of the statistically significant biometric responses for each sample per culture. Different letters denote significant differences based on ANOVA and Fisher's least significant difference post hoc test ($\alpha = 0.05$).

Author Contributions: Conceptualization, S.F. and C.G.V.; Data curation, S.F. and C.G.V.; Formal analysis, S.F. and C.G.V.; Investigation, S.F. and C.G.V.; Methodology, Y.Y.W. and C.G.V.; Project administration, S.F. and C.G.V.; Software, S.F. and C.G.V.; Supervision, S.F. and C.G.V.; Validation, S.F. and C.G.V.; Visualization, S.F. and C.G.V.; Writing—original draft, S.F. and C.G.V.; Writing—review and editing, S.F., Y.Y.W. and C.G. All authors have read and agreed to the published version of the manuscript.

Funding: This research received no external funding.

Conflicts of Interest: The authors declare no conflict of interest.

References

1. House, J. Consumer acceptance of insect-based foods in the Netherlands: Academic and commercial implications. *Appetite* **2016**, *107*, 47–58. [CrossRef] [PubMed]
2. La Barbera, F.; Verneau, F.; Amato, M.; Grunert, K. Understanding Westerners' disgust for the eating of insects: The role of food neophobia and implicit associations. *Food Qual. Prefer.* **2018**, *64*, 120–125. [CrossRef]
3. Hartmann, C.; Siegrist, M. Insects as food: Perception and acceptance. Findings from current research. *Ernahr. Umsch.* **2017**, *64*, 44–50.
4. Lensvelt, E.J.; Steenbekkers, L. Exploring consumer acceptance of entomophagy: A survey and experiment in Australia and the Netherlands. *Ecol. Food. Nutr.* **2014**, *53*, 543–561. [CrossRef]
5. Wilkinson, K.; Muhlhausler, B.; Motley, C.; Crump, A.; Bray, H.; Ankeny, R. Australian consumers' awareness and acceptance of insects as food. *Insects* **2018**, *9*, 44. [CrossRef]
6. van Huis, A.; Itterbeeck, J.V.; Klunder, H.; Mertens, E.; Halloran, A.; Muir, G.; Vantomme, P. *Edible Insects: Future Prospects for Food and Feed Security*; FAO: Rome, Italy, 2013.
7. Rumpold, B.A.; Schlüter, O. Insect-based protein sources and their potential for human consumption: Nutritional composition and processing. *Anim. Front.* **2015**, *5*, 20–24.
8. Rumpold, B.A.; Schlüter, O.K. Nutritional composition and safety aspects of edible insects. *Mol. Nutr. Food Res.* **2013**, *57*, 802–823. [CrossRef]
9. Oonincx, D.G.; van Itterbeeck, J.; Heetkamp, M.J.; van den Brand, H.; van Loon, J.J.; van Huis, A. An exploration on greenhouse gas and ammonia production by insect species suitable for animal or human consumption. *PLoS ONE* **2010**, *5*, e14445. [CrossRef]
10. Kouřimská, L.; Adámková, A. Nutritional and sensory quality of edible insects. *NFS J.* **2016**, *4*, 22–26. [CrossRef]
11. Van Huis, A. Did early humans consume insects? *J. Insects Food Feed* **2017**, *3*, 161–163. [CrossRef]
12. Verbeke, W. Profiling consumers who are ready to adopt insects as a meat substitute in a Western society. *Food Qual. Prefer.* **2015**, *39*, 147–155. [CrossRef]
13. Sogari, G. Entomophagy and Italian consumers: An exploratory analysis. *Prog. Nutr.* **2015**, *17*, 311–316.

14. Gmuer, A.; Guth, J.N.; Hartmann, C.; Siegrist, M. Effects of the degree of processing of insect ingredients in snacks on expected emotional experiences and willingness to eat. *Food Qual. Prefer.* **2016**, *54*, 117–127. [CrossRef]

15. Viejo, G.C.; Fuentes, S.; Howell, K.; Torrico, D.; Dunshea, F. Robotics and computer vision techniques combined with non-invasive consumer biometrics to assess quality traits from beer foamability using machine learning: A potential for artificial intelligence applications. *Food Control.* **2018**, *92*, 72–79. [CrossRef]

16. Viejo, G.C.; Fuentes, S.; Howell, K.; Torrico, D.; Dunshea, F. Integration of non-invasive biometrics with sensory analysis techniques to assess acceptability of beer by consumers. *Physiol. Behav.* **2019**, *200*, 139–147. [CrossRef]

17. Gunaratne, N.M.; Fuentes, S.; Gunaratne, T.M.; Torrico, D.D.; Francis, C.; Ashman, H.; Viejo, C.G.; Dunshea, F.R. Effects of packaging design on sensory liking and willingness to purchase: A study using novel chocolate packaging. *Heliyon* **2019**, *5*, e01696. [CrossRef]

18. Megido, R.C.; Gierts, C.; Blecker, C.; Brostaux, Y.; Haubruge, É.; Alabi, T.; Francis, F. Consumer acceptance of insect-based alternative meat products in Western countries. *Food Qual. Prefer.* **2016**, *52*, 237–243. [CrossRef]

19. Smith, T.W. *The Book of Human Emotions: An Encyclopedia of Feeling from Anger to Wanderlust*; Profile Books: London, UK, 2015.

20. Torrico, D.D.; Fuentes, S.; Gonzalez Viejo, C.; Ashman, H.; Gunaratne, N.M.; Gunaratne, T.M.; Dunshea, F.R. Images and chocolate stimuli affect physiological and affective responses of consumers: A cross-cultural study. *Food Qual. Prefer.* **2018**, *65*, 60–71. [CrossRef]

21. Gunaratne, N.M.; Fuentes, S.; Gunaratne, T.M.; Torrico, D.D.; Ashman, H.; Francis, C.; Gonzalez Viejo, C.; Dunshea, F.R. Consumer acceptability, eye fixation, and physiological responses: A study of novel and familiar chocolate packaging designs using eye-tracking devices. *Foods* **2019**, *8*, 253. [CrossRef]

22. Schouteten, J.J.; De Steur, H.; de Pelsmaeker, S.; Lagast, S.; Juvinal, J.G.; de Bourdeaudhuij, I.; Verbeke, W.; Gellynck, X. Emotional and sensory profiling of insect-, plant-and meat-based burgers under blind, expected and informed conditions. *Food Qual. Prefer.* **2016**, *52*, 27–31. [CrossRef]

23. Castro, M.; Chambers, E. Consumer Avoidance of Insect Containing Foods: Primary Emotions, Perceptions and Sensory Characteristics Driving Consumers Considerations. *Foods* **2019**, *8*, 351. [CrossRef] [PubMed]

24. Jensen, N.H.; Lieberoth, A. We will eat disgusting foods together–Evidence of the normative basis of Western entomophagy-disgust from an insect tasting. *Food Qual. Prefer.* **2019**, *72*, 109–115. [CrossRef]

25. Jaeger, S.R.; Lee, S.M.; Kim, K.-O.; Chheang, S.L.; Jin, D.; Ares, G. Measurement of product emotions using emoji surveys: Case studies with tasted foods and beverages. *Food Qual. Prefer.* **2017**, *62*, 46–59. [CrossRef]

26. Kaye, L.K.; Wall, H.J.; Malone, S.A. "Turn that frown upside-down": A contextual account of emoticon usage on different virtual platforms. *Computs. Hum. Behav.* **2016**, *60*, 463–467. [CrossRef]

27. Jaeger, S.R.; Vidal, L.; Kam, K.; Ares, G. Can emoji be used as a direct method to measure emotional associations to food names? Preliminary investigations with consumers in USA and China. *Food Qual. Prefer.* **2017**, *56*, 38–48. [CrossRef]

28. Gunaratne, T.M.; Fuentes, S.; Gunaratne, N.M.; Torrico, D.D.; Gonzalez Viejo, C.; Dunshea, F.R. Physiological responses to basic tastes for sensory evaluation of chocolate using biometric techniques. *Foods* **2019**, *8*, 243. [CrossRef]

29. Gunaratne, M. Implementation of non-invasive biometrics to identify effects of chocolate packaging towards consumer emotional and sensory responses. Ph.D. Thesis, The University of Melbourne, Melbourne, Australia, 2019.

30. Fuentes, S.; Viejo, G.C.; Torrico, D.; Dunshea, F. Development of a biosensory computer application to assess physiological and emotional responses from sensory panelists. *Sensors* **2018**, *18*, 2958. [CrossRef]

31. Menozzi, D.; Sogari, G.; Veneziani, M.; Simoni, E.; Mora, C. Eating novel foods: An application of the Theory of Planned Behaviour to predict the consumption of an insect-based product. *Food Qual. Prefer.* **2017**, *59*, 27–34. [CrossRef]

32. Dibeklioglu, H.; Gevers, T. Automatic estimation of taste liking through facial expression dynamics. *IEEE Trans. Affect. Comput.* **2018**. [CrossRef]

33. Zhi, R.; Wan, J.; Zhang, D.; Li, W. Correlation between hedonic liking and facial expression measurement using dynamic affective response representation. *Food Res. Int.* **2018**, *108*, 237–245. [CrossRef]

34. Rocha-Parra, D.; García-Burgos, D.; Munsch, S.; Chirife, J.; Zamora, M.C. Application of hedonic dynamics using multiple-sip temporal-liking and facial expression for evaluation of a new beverage. *Food Qual. Prefer.* **2016**, *52*, 153–159. [CrossRef]
35. Buss, D. Food Companies Get Smart about Artificial Intelligence. *Food Techinol.* **2018**, *72*, 26–41.
36. Gonzalez Viejo, C.; Fuentes, S. Beer Aroma and Quality Traits Assessment Using Artificial Intelligence. *Fermentation* **2020**, *6*, 56. [CrossRef]
37. Hetherington, M.; Madrelle, J.; Nekitsing, C.; Barends, C.; de Graaf, C.; Morgan, S.; Parrott, H.; Weenen, H. Developing a novel tool to assess liking and wanting in infants at the time of complementary feeding–The Feeding Infants: Behaviour and Facial Expression Coding System (FIBFECS). *Food Qual. Prefer.* **2016**, *48*, 238–250. [CrossRef]
38. Nekitsing, C.; Madrelle, J.; Barends, C.; de Graaf, C.; Parrott, H.; Morgan, S.; Weenen, H.; Hetherington, M. Application and validation of the Feeding Infants: Behaviour and Facial Expression Coding System (FIBFECS) to assess liking and wanting in infants at the time of complementary feeding. *Food Qual. Prefer.* **2016**, *48*, 228–237. [CrossRef]
39. Torrico, D.D.; Fuentes, S.; Viejo, C.G.; Ashman, H.; Gurr, P.A.; Dunshea, F.R. Analysis of thermochromic label elements and colour transitions using sensory acceptability and eye tracking techniques. *LWT* **2018**, *89*, 475–481. [CrossRef]
40. Viejo, G.C.; Torrico, D.; Dunshea, F.; Fuentes, S. Development of Artificial Neural Network Models to Assess Beer Acceptability Based on Sensory Properties Using a Robotic Pourer: A Comparative Model Approach to Achieve an Artificial Intelligence System. *Beverages* **2019**, *5*, 33. [CrossRef]

Article

Reading Food Experiences from the Face: Effects of Familiarity and Branding of Soy Sauce on Facial Expressions and Video-Based RPPG Heart Rate

Rene A. de Wijk [1,*], Shota Ushiama [1,2], Meeke Ummels [1], Patrick Zimmerman [3], Daisuke Kaneko [2] and Monique H. Vingerhoeds [1]

[1] Wageningen Food & Biobased Research Institute, Wageningen University & Research, 6700 AA Wageningen, The Netherlands; s.ushiama@kikkoman.nl (S.U.); meeke.ummels@wur.nl (M.U.); monique.vingerhoeds@wur.nl (M.H.V.)

[2] Kikkoman Europe R&D Laboratory B.V., 6709 PA Wageningen, The Netherlands; d.kaneko@kikkoman.nl

[3] Noldus Information Technology, 6709 PA Wageningen, The Netherlands; Patrick.Zimmerman@noldus.nl

* Correspondence: rene.dewijk@wur.nl

Abstract: Food experiences are not only driven by the food's intrinsic properties, such as its taste, texture, and aroma, but also by extrinsic properties such as visual brand information and the consumers' previous experiences with the foods. Recent developments in automated facial expression analysis and heart rate detection based on skin color changes (remote photoplethysmography or RPPG) allow for the monitoring of food experiences based on video images of the face. RPPG offers the possibility of large-scale non-laboratory and web-based testing of food products. In this study, results from the video-based analysis were compared to the more conventional tests (scores of valence and arousal using Emojis and photoplethysmography heart rate (PPG)). Forty participants with varying degrees of familiarity with soy sauce were presented with samples of rice and three commercial soy sauces with and without brand information. The results showed that (1) liking and arousal were affected primarily by the specific tastes, but not by branding and familiarity. In contrast, facial expressions were affected by branding and familiarity, and to a lesser degree by specific tastes. (2) RPPG heart rate and PPG both showed effects of branding and familiarity. However, RPPG heart rate needs further development because it underestimated the heart rate compared to PPG and was less sensitive to changes over time and with activity (viewing of brand information and tasting). In conclusion, this study suggests that recording of facial expressions and heart rates may no longer be limited to laboratories but can be done remotely using video images, which offers opportunities for large-scale testing in consumer science.

Keywords: RPPG and PPG heart rate; facial expressions; emojis; branding; familiarity; soy sauce

Citation: de Wijk, R.A.; Ushiama, S.; Ummels, M.; Zimmerman, P.; Kaneko, D.; Vingerhoeds, M.H. Reading Food Experiences from the Face: Effects of Familiarity and Branding of Soy Sauce on Facial Expressions and Video-Based RPPG Heart Rate. *Foods* **2021**, *10*, 1345. https://doi.org/10.3390/foods10061345

Academic Editor: Damir Torrico

Received: 18 May 2021
Accepted: 9 June 2021
Published: 10 June 2021

Publisher's Note: MDPI stays neutral with regard to jurisdictional claims in published maps and institutional affiliations.

check for **updates**

1. Introduction

Food perception typically starts before the food is placed in the mouth. Before the food is consumed, consumers typically smell the food and see the food, often with the food package with brand name. Consequently, prior to tasting, consumers will already have expectations regarding the food's taste and flavor based on visual, smell, tactile, and sometimes even auditory cues. These expectations affect whether the consumer will eat the food. Numerous studies have focused especially on visual cues. They have demonstrated the effects of branding, names and sensory descriptors, and health and ingredient labels on food choice decisions. These "extrinsic" variables provide consumers with information regarding the product's quality, which is part of "brand equity" [1,2]. Brand equity combines perceived quality with factors such as brand loyalty, name awareness, and other associations that the consumer may have with the brand [3]. Obviously, the factors vary with the consumer's experience with the product and aid consumers in selecting products that fit their attitudes [4,5]. Extrinsic variables may also exert direct effects on taste experiences [6,7].

Models such as the assimilation and contrast model [6,8] describe various ways of how expectations interact with actual experiences, depending on their overlap and discrepancies. In the assimilation model, the discrepancy is relatively small, and experiences are adjusted to the expectations. In contrast, when the discrepancy is relatively large, the differences between expectations and experiences are amplified. As a result, ratings for the experience will shift in the opposite direction of the expectations. Recently, examples of assimilation and contrast were provided by Głuchowski et al. [9], who investigated emotional and hedonic reactions to traditional and new innovative food dishes. The results showed that for traditional dishes, expected liking based on visual aspects was in line with experienced liking during tasting, whereas for new innovative dishes the experienced liking fell short compared to the expected liking. The results also demonstrate the importance of taking consumers' characteristics such as food neophobia and consumer innovativeness into account.

Most studies used subjective ratings and questionnaires, so-called explicit measures, to assess the effects of expectations on taste experiences, i.e., these studies rely on introspection while consumers may not even be aware of the way expectations affect their experiences (e.g., [10]). Additionally, virtually all studies measured the effect of expectation indirectly by measuring its effect on subsequent taste experiences, i.e., only very few studies try to measure the expectations themselves, except for Thomson, who uses a specially developed technique [11].

On the other hand, implicit measures of autonomic nervous system (ANS) responses, such as heart rate, and their behavioral correlates, such as facial expressions, offer a promising alternative. In fact, these measurements (1) do not rely on introspection and offer insight into the subconscious as well as conscious processes and (2) can be measured continuously during expectations and experiences (e.g., [12]). The continuous measurement of ANS responses and facial expressions facilitate insights in the different stages of the temporal development of responses and have been corner stones of theories of emotions, such as appraisal theories. Early responses reflect aspects of stimuli, such as the stimulus' novelty, relevance, or pleasantness. Later responses include cognitive reactions on the earlier responses.

Despite their potential, facial expressions and ANS responses have been used only in relative few food studies and with varying success: some studies failed to find effects on ANS responses or facial expressions between the tastes of foods, such as chocolates [13], beers [14], and energy drinks [15], whereas others found significant effects, even though they were small and difficult to interpret (e.g., [16] for juices, [17] for beers, and [18] for breakfast drinks). In studies where the foods were not only tasted but also visually inspected, the results typically showed clearer differences between foods. In a recent study, Bercik et al. [19] measured facial expressions as well as eye movements for three dishes with the same ingredients but served in three different ways. Results showed that the way the food was visually presented had significant emotional effects and influenced the way participants visually inspected the foods. Similarly, facial expressions were intensified when foods were tested in a natural consumption environment, such as one's own home, rather than in the unrealistic and uninspiring setting of a sensory laboratory [20], which suggests that this type of measure reflects the complete food experience determined by a combination of variables such as the food, the physical location, and the social environment, rather than by specific taste experiences.

Unfortunately, ANS measurements and some facial expression measurements typically cannot be used in realistic non-laboratory settings because they require sensors attached to face and body [9,13,14,16,18,21–24]. Recent technical developments have resulted in video-based automated analysis of facial expressions and even more recently in video-based recordings of heart rate using remote photoplethysmography or RPPG (e.g., [20,25]). If these new methods prove to be a good alternative for the gold-standard methods such as electrocardiogram (ECG) and photoplethysmography (PPG), then these measurements would no longer require physical contact between the sensor and human volunteer. This would open a whole new range of applications such as testing in realistic consumption environments, and internet testing using a webcam, which is built-in in most notebooks nowadays.

This study explored the use of video-based facial expressions, RPPG heart rate, and Emoji scores of valence and arousal [26], for branded and unbranded foods. The results of this study complement results from the same study reported previously. Those results, based on invasive PPG heart rates, as well as skin conductance measures, showed that facial expressions, PPG heart rate, and skin conductance varied significantly with the level of familiarity of product for the consumer and with the type of branding and varied less with the brand of soy sauce. Visual analogue scale (VAS) liking and arousal scores also varied with the level of familiarity and with the brand of soy sauce. In contrast with the other measures, VAS scores did not vary with branding [27].

This study aims at (1) the comparison of valence and arousal from video-based facial expressions with Emoji scores during food tasting, and (2) the comparison of video-based RPPG heart rate and PPG heart rate during food viewing and tasting. Specifically, the sensitivity of the methods to detecting effects of branding, soy sauce, and degree of consumers' familiarity will be compared. For these purposes, samples of branded and unbranded soy sauces are presented to experienced and inexperienced consumers. During visual inspection and tasting, video-based facial expressions, RPPG and PPG heart rate, and Emoji scores of valence and arousal [26] were recorded.

2. Materials and Methods

2.1. Participants

Forty healthy participants (30 females and 10 males aged between 20 and 59 yrs. with an average age of 43 yrs.) were recruited from the consumer database of Food, Health, and Consumer Research (Wageningen Food and Biobased Research) after screening with an online survey (EyeQuestion Software, Logic8 B.V., The Netherlands). The database consists of 2945 persons (1015 males and 1930 females all living in or near Wageningen and aged between 19 and 95 yrs). Exclusion criteria were allergies or intolerance to wheat, gluten, soybean, or rice allergy. All participants provided written consent to participate in the experiment. In the recruitment questionnaire, participants were asked to indicate the frequency of their soy sauce use using an 8-point scale (more than three times a week, 1–2 times a week, once a month, once per 3–4 month, 1–2 times a year, less than once a year, never used, and I do not know if I use soy sauce) and daily use soy sauce brand. The participants were assigned to one of two groups depending on their frequencies of use. Twenty-two participants, 17 females and 5 males, who used soy sauce either never or maximum once per month were assigned to the low-frequency user group. Eighteen participants, 15 females and 3 males, who used soy sauce at least once per week were assigned to the high-frequency user group.

The participants received financial compensation of 10 euro after finalizing the study. The study was approved by the Social Ethical Committee of Wageningen University and Research.

2.2. Soy Sauces

Three soy sauces were selected based on the similarity of ingredients and availability in the Netherlands. Soy sauce is a condiment with a distinctive flavor that is familiar to most Asian consumers, and nowadays, a large variety of soy sauces are increasingly being offered in regular supermarkets to the mainstream consumers. Despite its growing popularity, soy sauce is still consumed regularly by a relatively small group of consumers, while most consumers either never consume soy sauce or very infrequently [28]. Recently, the frequency of use of soy sauce was related to factors such as food neophobia and preferences for soy sauces in general as well as preferences for specific brands of soy sauce [29]. Based on this study, two familiar brands of soy sauces (Kikkoman (Kikkoman Foods Europe B.V., Sappemeer, The Netherlands) and Inproba (Inproba B.V., Baarn, The Netherlands)) were selected that shared the typical savory salty taste of soy sauce and one unfamiliar brand of soy sauce with a clearly different taste (Maekrua Gold Label® Soya Sauce Formula 1 (Mae Krua Sri Ruen Company Limited, Bangkok, Thailand)). In order to study the effect of brand familiarity, the familiar Inproba soy sauce was presented in an unfamiliar bottle (Wan Ja Shan from Taiwan). Samples of the soy sauces (1 mL, either with or without branding information) were mixed together with 16 g of shaped Sticky-rice (Yumenishiki, JFC Deutschland GmBH, Düsseldorf, Germany) in a styrofoam cup and presented at a comfortable temperature. White rice with soy sauce was selected as a carrier because it is a relatively common way to use soy sauce in the Netherlands. Rice had been cooked in a commercial rice cooker and kept warm at 60 °C in the incubator and portioned before each session.

2.3. Procedure

Participants participated in one 1-h session, which started with an explanation of the procedure, followed by the attachment of the PPG heart rate sensor on the fingers of the non-dominant hand, and by two practice trials. Participants were seated in front of a laptop with a webcam (Microsoft LifeCam) that recorded their face. Instructions and images of the unbranded or branded soy sauces were shown on the laptop screen using E-Prime software (E-Prime, Science Plus Group, Groningen, The Netherlands). Participants were instructed that their perception of soy sauces was studied, and that to avoid unnecessary movements that may interfere with the measurements, small samples of rice with soy sauce would be put in their mouth by the experimenter. The sample should be chewed normally, and the participant was instructed to signal the swallow with a hand signal, after which scores were entered by a mouse click.

Next, participants received three blocks of stimulus trials in which he/she received taste samples of cooked rice mixed one of three soy sauces. In each trial, participants, either familiar or unfamiliar with soy sauces, first received an auditory warning signal followed after 5 s by an image (*viewing phase*) for 5 s showing either a generic volume of soy sauce in a white cup (unbranded block 1), or the same white cup with soy sauce combined with the bottle of one of the soy sauces showing the package of the product (blocks 2 and 3, with matched branding and non-matched branding, respectively). After the 5 s viewing period, the image disappeared during an anticipatory waiting period of approximately 5 s. Next, the image reappeared and the participants were given a spoon of the mixture of rice and soy sauce in the mouth by the experimenter who simultaneously pressed the enter key of the keyboard to signal the start of tasting (*tasting phase*) in the data recordings. In the branded match block 2, the brand shown in the image and the tasted soy sauce were the same, whereas they were different in the branded non-match block 3 (results will not be reported in here). After each sample, which was chewed typically for at least 10 s, the participant first scored the liking of the sample using a visual analogue scale (VAS), followed by a just-right scale for saltines. Next, participants indicated the perceived valence and arousal using a so-called "EmojiGrid" [26,30] (see Figure 1). Only the Emoji scores of valence and arousal will be reported here. Heart rate responses and facial expressions were recorded during each of the phases in all trials. Each of the three soy sauces was presented twice in block 1 and four times in block 2 in a randomized order. To avoid possible unwanted interference of the non-matched condition with the other conditions, the non-matched condition (not reported here) was tested last in block 3 with only two of the three soy sauces presented once. This resulted in a total number of 20 trials presented at a rate of approximately 1 trial per 150 s. The order of the blocks was the same for all participants. The procedures are shown schematically in Scheme 1.

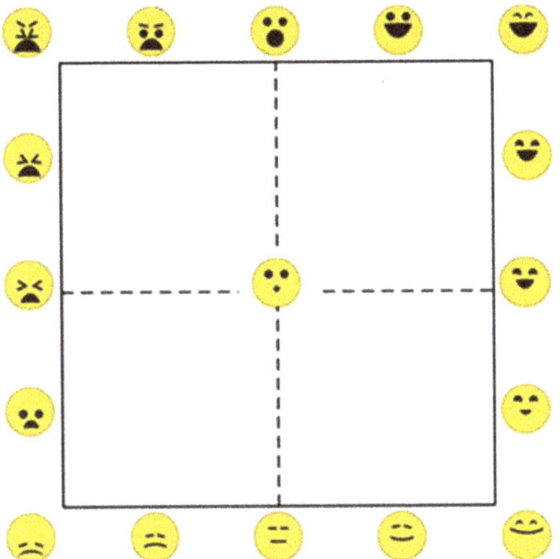

Figure 1. The EmojiGrid used in the study to measure valence and arousal [25].

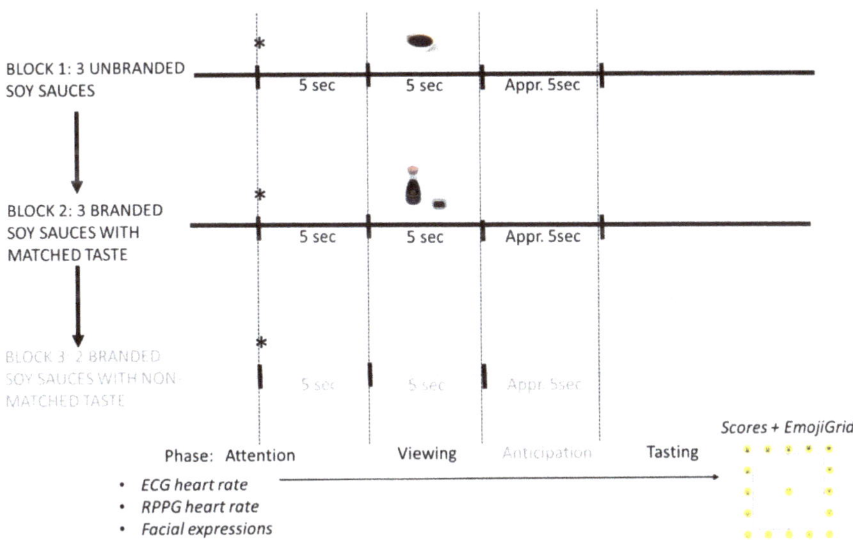

Scheme 1. Schematic representation of the study procedure. Participants were presented with 3 blocks of trials. Each trial started with an attention signal (asterisk), followed by an image of branded or unbranded image soy sauce, a viewing/anticipation period, and a period where rice with soy sauce was chewed. ECG and RPPG heart rate, skin conductance, and facial expressions were recorded continuously. VAS and EmojiGrid scores were recorded at the end of each trial. Heart rates, facial expressions, and Emoji scores for block 1 and 2 are reported in this manuscript.

2.4. Physiological ANS Measures

PPG heart rate was collected using a BIOPAC MP150 system and AcqKnowledge 5.0.4 data acquisition software (BIOPAC Systems Inc., Santa Barbara, CA, USA). Heart rate was measured using the TSD200 PPG transducer attached to the index finger of the non-dominant hand. The TSD200 transducer was also connected to the PPG/EDA transmitter. Data were stored on a desktop PC (sample rate 250 Hz). The PPG signal was processed real-time in AcqKnowledge using the Pulse Rate calculation channel, resulting in the pulse/heart rate in beats-per-minute (bpm).

RPPG heart rate was recorded remotely based on photoplethysmography (RPPG), a technique that measures the small changes in color caused by changes in blood volume under the skin epidermis using a multi-wavelength RGB camera [31,32]. Pilot food studies in which RPPG heart rate was compared to the gold-standard ECG and PPG during chewing and not-chewing showed that RPPG heart rate was generally approximately 6 BPM lower than ECG and PPG heart rate, especially during chewing. Overall, the correspondence between RPPG and ECG/PPG heart rate was satisfactory [33].

Heart rate in combination with other physiological parameters have previously been related to food preferences and emotions (see [12]).

2.5. Facial Expressions

Video segments of the consumption of each bite were stored together with the participant's code, the product code, and time and date information. Facial expression data were automatically analyzed per time frame of 0.04 s by FaceReader 8.0 (Noldus Information Technology, Wageningen, The Netherlands) in three steps. The face is detected in the first step using the Viola-Jones algorithm [34]. Next, the face is accurately modelled using an algorithmic approach [35]. Based on the Active Appearance method described by Cootes and Taylor [36], the model is trained with a database of annotated images that describes over 500 key points in the face and the facial texture of the face. Finally, the actual classification of the facial expressions is based on an artificial neural network trained with 10,000 manually annotated images. The face classification provides the output of seven basic expressions (happy, sad, angry, surprised, scared, disgusted, and contemptuous), one neutral state on the basis of the Facial Action Coding System developed by Ekman and Friesen [37], three "affective attitudes" (interest, boredom, and confusion), and arousal and valence dimensions based on combinations of facial expressions. Valence scores were calculated per time frame from the FaceReader happiness score minus the most intense FaceReader negative emotion score (sad, angry, surprised, scared, disgusted, and contemptuous). FaceReader scores for each emotional expression, except arousal, range from 0 (emotion is not detected) to 1 (maximal detection) and are based on intensity judgments of human experts. Arousal scores range from −1 to 1. FaceReader allows for the simultaneous presence of multiple emotions. Only the arousal and valence dimensions were used for this study.

FaceReader was validated by others using the Radboud Faces Database, a standardized test with images of expressions associated with basic emotions. The test persons in the images were trained to pose a particular emotion and the images were labelled accordingly by the researchers. Subsequently, the images were analyzed in FaceReader. Accuracy of the assessment of the emotions by FaceReader varied between 84.4% for scared and 95.9% for happy, with an average of 90% [38]. Other validation studies showed superior performance of FaceReader for neutral faces (90% correct recognition for FaceReader versus 59% for humans) [39].

A more detailed description of the science behind FaceReader can be found at http://info.noldus.com/free-white-paper-on-facereader-methodology/ (accessed on 18 May 2021). Whether all possible emotional expressions can be categorized by these six emotions, affective attitudes, and arousal/valence dimensions remains a matter of debate (e.g., [40]).

2.6. Data Analysis

Physiological responses and facial expressions. Facial expressions and heart rates were collected during the five-second intervals of the viewing, anticipation, and tasting phases. The duration of the anticipation phase was not exactly controlled because it ended when the participant was manually fed the test food and started chewing, and only the results of the viewing and tasting phases will be reported here. For each phase, averages per second were used for further analysis. PPG heart rate was not successfully collected for five participants. To allow the proper comparison, RPPG heart rates of these participants were also omitted from further analysis. For the remaining participants, heart rates were missing in 4% of all cases, and facial expressions were missing in 1% of all cases.

EmojiGrid scores. The positions of the scores on the EmojiGrid valence (X) and arousal (Y) axis were converted into numerical values (1–10).

Results were analyzed with Mixed Model Anovas (IBM® SPSS® statistics, version 25, Armonk, New York, NY, USA) with the participant as a random factor, and with the familiarity with soy sauce (familiar and unfamiliar users), soy sauce (3), duration (5 s), and condition (unbranded and branded) as fixed factors. The usage of soy sauce was a between-subject factor, and the others were within-subject factors. LSD post-hoc tests were used for main effects. 95% Confidence intervals were used for post-hoc tests of interaction effects.

3. Results

3.1. General

The results of facial expressions and RPPG/PPG heart rates will be presented for the viewing and tasting phases, together with Emoji valence and arousal scores, which were collected only in the tasting phase. The effects of each phase will be presented, as well as the effects of soy sauce, familiarity with soy sauce, and branding (no branding, matched branding). Post-hoc test results will only be reported when they are significant.

3.2. Viewing Phase: Facial Expressions

Facial expressions of valence. Facial expressions to images of foods became more negatively valenced when branding information was added ($F(1,1092) = 27.2$, $p < 0.001$). The effects of branding varied with the type of soy sauce (interaction $F(2,1092) = 3.9$, $p = 0.02$). Facial expressions did not vary with time.

Facial expressions of arousal. Facial expressions related to arousal became intensified with branding information ($F(1,1092) = 35.4$, $p < 0.001$). Facial expressions varied overall between soy sauces ($F(2, 1091) = 3.4$, $p = 0.03$), and varied with combination of soy sauce and experience (interaction: $F(2,1091) = 4.5$, $p = 0.01$). Branding affected familiar and unfamiliar participants differently (interaction: $F(1,1091) = 4.3$, $p = 0.04$). Facial expressions did not vary with time. Results are shown in Figure 2.

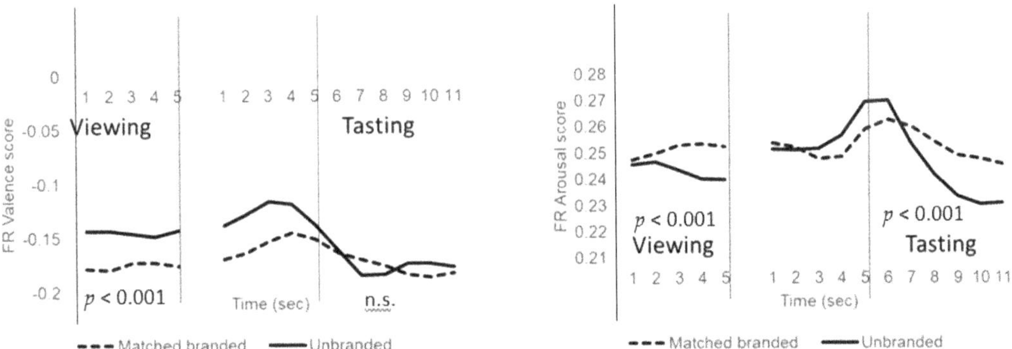

Figure 2. Facial expressions related to valence (left) and arousal (right) during 5 s of viewing, 5 s of anticipation between viewing and tasting, and 6 s of tasting (6–11 s) of unbranded (solid line) and matched branded (dashed line) stimuli. Results were averaged across participants and stimuli. Significance levels of branded/unbranded differences are indicated by *p*-values. N.s. = not significant.

3.3. Viewing Phase: RPPG and ECG Heart Rate

RPPG heart rate. RPPG heart rates during viewing of the food lowered from 59.7 to 57.9 BPM when branding information was also presented (F(1,1080) = 24.0, $p < 0.001$). RPPG heart rates varied between soy sauces (F(2,1080) = 3.2, $p = 0.04$), and with combinations of soy sauce, familiarity, and branding (F(2,1080) = 5.5, $p < 0.01$). RPPG heart rate did not vary with time.

PPG heart rate. PPG heart rates during viewing were 9–10 BPM higher than RPPG heart rates but showed a similar decrease with branding information (from 70.6 to 68.6 BPM, F(1,957) = 78.6, $p < 0.001$). The effect of branding varied with the participants' familiarity (F(1,957) = 10.2, $p = 0.001$). The lowering of PPG heart rate with branding information was larger for unfamiliar participants. PPG heart rate did not vary with time. Results are shown in Figure 3.

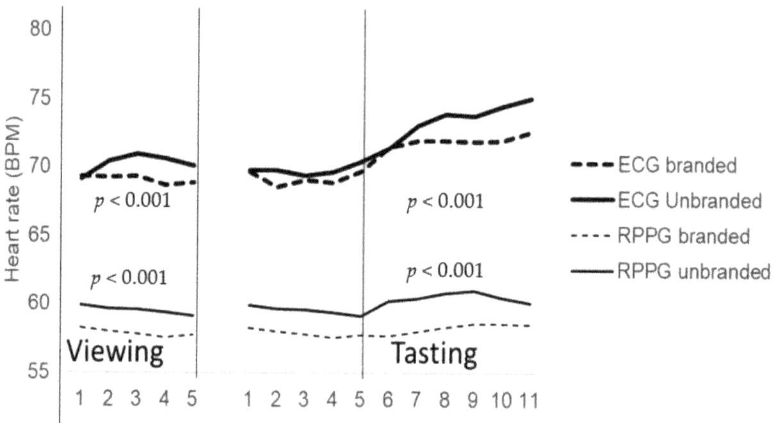

Figure 3. Heart rates measured by ECG and RPPG during 5 s of viewing, 5 s of anticipation between viewing and tasting, and 6 s of tasting (6–11 s) of unbranded (solid line) and matched branded (dashed line) stimuli. Significance levels of branded/unbranded differences are indicated by p-values.

3.4. Tasting Phase: Facial Expressions

Facial expressions of valence. Facial expressions became increasing more negatively valenced during tasting (effect of time: F(4,1099) = 7.4, $p < 0.001$). Brand information was associated with more negatively valenced facial expressions in low-frequency users, whereas branding did not affect facial expressions of high-frequency users (interaction: F(1,1099) = 5.8, $p = 0.02$).

Facial expressions of arousal. Overall, arousal was more intense at the beginning of tasting and became less intense there after (F(4,1098) = 27.9, $p < 0.001$). Facial expressions during tasting became more aroused with branding information (F(1,1098) = 27.2, $p < 0.001$). The branding effect was strongest for inexperienced participants (F(1,1098) = 5.5, $p = 0.02$). Expressions of experienced participants were most aroused by Kikkoman, whereas those of inexperienced participants were most aroused by Inproba (interaction F(2,1098) = 3.3, $p = 0.04$). Results are shown in Figure 2.

3.5. Tasting Phase: RPPG and PPG Heart Rate

RPPG heart rate. RPPG heart rates during tasting lowered from 60.2 to 57.8 BPM with branding information (F(1,1074) = 44.5, $p < 0.001$). This decrease was especially large for inexperienced participants (interaction: F(1,1074) = 7.6, $p < 0.001$). RPPG heart rates of experienced participants were highest for Kikkoman, whereas those for inexperienced participants were highest for Formula 1 (interaction: F(2,1073) = 13.7, $p < 0.001$).

PPG heart rate. PPG heart rates during tasting were 12–14 BPM higher than RPPG heart rates, but also showed a lowered PPG heart rate (from 74.6 to 71.9 BPM) with branding information (F(1,954) = 116.2, $p < 0.001$). PPG heart rates varied with soy sauce (F(2,954) = 5.2, $p < 0.001$). PPG heart rates increased during chewing from 71.6 to 74.6 BPM (F(4,954) = 17.2, $p < 0.001$). This increase was especially large for inexperienced participants (F(4,954) = 2.8, $p = 0.02$). Results are shown in Figure 3.

3.6. Explicit Valence and Arousal during Tasting

Emoji liking (EmojiX scores) varied significantly with soy sauce (F(2,190) = 6.2, $p < 0.01$). Kikkoman was liked best (6.1), followed by Formula1 (5.7) and Inproba (5.3). Inclusion of (matched) branding information or familiarity group did not affect Emoji liking scores. Emoji arousal (EmojiX scores) also varied significantly with soy sauce F(1,190) = 4.5, $p = 0.01$. Arousal was strongest for Kikkoman (3.2) and Formula1 (3.2) and weakest for Inproba (2.70). Inclusion of (matched) branding information or familiarity group did not affect Emoji arousal scores.

4. Discussion

The study investigated the effects of consumer' experience and branding on reactions to the sight and taste of branded and unbranded soy sauces. This study aimed at (1) the comparison of valence and arousal from video-based facial expressions and from Emoji scores during food tasting, and (2) the comparison of video based RPPG heart rate and PPG heart rate during food viewing and tasting.

With regard to the first aim, the study showed differences between the results from implicit (facial expressions and heart rate) and explicit (scores) responses. Explicit Emoji taste scores of arousal and valence showed significant differences between the brands of soy sauce but showed no effects of branding and consumer' experience. In general, facial expressions of arousal and valence primarily showed effects of branding and consumer' experiences. Moreover, these differences were not only found during tasting but also prior to tasting when the participants' looked at the unbranded and branded foods. This illustrates the importance of consumers' expectations based on so-called extrinsic food factors, such as the food's visible appearance, the packaging, and branding information. It also suggests that measures such as facial expressions and heart rate are better suited to investigate the role of extrinsic factors than the more traditional measures such as scores. Overall, this outcome corresponds with other results of the same study reported earlier [27], as well as results of a recent study that also showed that scores reflected differences between brands of soy sauce but were not affected by branding [29].

The PPG and facial expression results from the same study reported previously [27] failed to show systematic differences between soy sauces, whereas in the results reported here occasional differences between soy sauces were found during viewing as well as tasting. These differences are probably related to differences in the way the results of the earlier and this study were analyzed. In the previous study, changes in heart rate, and facial expressions, relative to the start of viewing and tasting were analyzed, whereas in this study absolute values were used. The fact that differences between soy sauces were found in the absolute values and not in the relative values suggest that these differences do not develop during tasting but have already developed previously during the viewing of the food and branding information, and the following anticipation phases, and are continued during tasting. This illustrates how measures such as facial expressions and heart rate provide a window into processes that start prior to tasting with expectations based on visual (as well as possibly olfactory and auditory) information and that affect experiences during subsequent tasting. These interactions between expectations and actual taste experiences are not (always) reflected in scores. The richness of the processing during visual inspection and subsequent tasting that is reflected in the heart rate and facial expressions may also be a reason why heart rates, and responses of the autonomic system in general, do not always show differences between products when they are based solely on taste responses. For example, Gunaratne [41] presented participants with unbranded chocolate samples and failed to find effects of specific chocolate tastes on heart rate and skin temperature (see also [13,14], who showed similar results for chocolate and beer, respectively). Other studies did demonstrate significant product effects on ANS responses, but these effects were typically small and difficult to interpret (e.g., [18]). The higher sensitivity of heart rate and facial expressions compared to explicit scores for factors such as branding that are not directly related to the food's taste properties is in line with our previous study. In that study, participants evaluated foods repeatedly in their own home and in the sensory laboratory [20]. Factors such as time of the day, day of the week, and sample preparation and presentation procedure were kept constant, whereas the physical location was systematically varied. Again, the results showed that explicit scores reflected primarily the differences between the test foods, whereas the heart rate and facial expressions reflected differences between test locations as well. Combined, the results of these studies suggest that physiological measures and facial expressions may reflect experiences in general and not only experiences directly related to foods. These experiences may be related to a specific mood that is affected by the food that is consumed and the circumstances in which the food is consumed. A close link between the consumption context and how consumers feel has been postulated by others [42], and these feelings are thought to be underlying modulators of food perception, food liking, and overall enjoyment of human eating experiences [43]. Others, such as Meiselman [44], have also stressed the importance of assessing foods and drinks in accurate contexts in order to increase the external validity of the results of consumer tests. The present study demonstrates that physiological measures and facial expressions may be especially well suited for the assessment of general food experiences.

Subjective rating scores such as the Emoji scores used in this study are single point measurements where all the consumer' processes during viewing, anticipation, and tasting are collapsed over time into one single score. An obvious result is that many of the nuances pre- and during tasting are blunted and possibly lost. Facial expressions and heart rate offer a continuous measurement of the pre- and during tasting processes. The fact that facial expressions and PPG heart rate vary between viewing and tasting, and show variation over time, illustrates that these consumer' processes are highly dynamical. In order to understand consumer' processes, these dynamics need to be considered when selecting the appropriate measurement. This means that the measurements need to be continuous, like facial expressions and heart rate, and not static like scores.

The second aim of the study was the comparison of heart rate recorded with non-invasive video-based RPPG and with invasive PPG (which relates closely to ECG, the ground truth). A successful comparison would mean that physiological heart rate recordings are no longer limited to laboratory studies but could be also applied to studies in realistic consumer situations. This is important because consumer' reactions to foods are not only driven by food properties but also by the physical and social context in which the food is consumed. Another possible new application would be internet-based surveys where large numbers of consumers can be reached. Unfortunately, the comparison of RPPG and PPG heart rates indicates that at this moment the RPPG method cannot fully replace the PPG and other invasive methods. RPPG heart rates were systematically approximately 10–12 BPM slower than PPG heart rates, and this difference seemed to be larger for higher PPG heart rates associated with for example chewing activity. Similar results were reported previously by Kanemura et al. [33], who speculated that some of the differences could be related to the effects of chewing activity on skin color. In addition, PPG heart rates changed rapidly over time during viewing and tasting, whereas RPPG heart rates were relatively stable. Rapid changes in PPG heart rates have been associated with orienting responses to new and unexpected stimuli and attentional processes (e.g., [21,45]). The results of this study suggest that the current RPPG heart rate measurements cannot (yet) be used to study this detailed level of processes. Even though RPPG heart rates based on video images currently lack the high level of sensitivity provided by "gold-standard" measurements, RPPG heart rates showed overall effects of consumers' experience, phase (viewing, anticipation, and tasting) and occasionally of brand of soy sauce, similar to PPG heart rates. Obviously, further development of the RPPG technique is needed for a closer match with PPG heart rates, and additional research is needed to explore the relation between both types of heart rate measurements further.

This research contributes—together with other studies—to the understanding of how experiences and expectations from packaging and branding interact with subsequent taste experiences. The results clearly demonstrate that taste reactions are directly related to reactions to packages and branding, and that both need to be considered for the development of new products. In traditional sensory research, taste experiences are typically tested with unbranded products, i.e., without any packaging and branding information. In subsequent consumer tests, only products that successfully pass the initial sensory taste test were tested together with branding and packaging information. This, as well as other research, indicate that this step-wise approach may not be the most efficient way to develop new successful products (see also [46]). This study also demonstrates the need to not only assess effects of branding/packaging on taste reactions but also to measure reactions to the branding/packaging itself. There reactions are especially difficult for consumers to articulate because they are often unaware of them. In those cases, measures such as facial expressions and physiological measurements may provide important insights, not only into the effects of branding/packaging but also into the interactions with specific consumer characteristics, such as the consumers' experience with the brand or product.

In conclusion, this study suggests that video-based measurements of facial expressions and RPPG heart rates offer opportunities for large-scale web-studies to investigate consumers' reactions to extrinsic food factors such as packages, labels, logos, and product information. In addition, facial expressions and RPPG heart rates may be well suited to study the effects of consumers' traits, such as level of experience with the foods, on their reactions to extrinsic food factors.

Author Contributions: Conceptualization: R.A.d.W., S.U., and D.K., Methodology: P.Z., R.A.d.W. Formal analysis: M.U., P.Z. Investigation, M.U., P.Z., and S.U. Writing original manuscript: R.A.d.W., D.K., and S.U., writing review + editing: M.H.V., P.Z., S.U., and M.U., supervision: M.H.V., project administration: M.H.V. All authors have read and agreed to the published version of the manuscript.

Funding: This study was partially funded by a grant from Kikkoman Europe R&D Laboratory B.V.

Institutional Review Board Statement: The study was conducted according to the guidelines of the Declaration of Helsinki and approved by the Social Sciences Ethics Committee (SEC). The Committee has concluded that the proposal deals with ethical issues in a satisfactory way and that it complies with the Netherlands Code of Conduct for Research integrity and the Netherlands Code of Ethics for Research in the Social and Behavioral Sciences involving Human Participants.

Informed Consent Statement: Informed consent was obtained from all subjects involved in the study.

Data Availability Statement: The data presented in this study are available on request from the corresponding author.

Conflicts of Interest: This study was partially funded by a grant from Kikkoman Europe R&D Laboratory B.V. Shota Ushiama and Daisuke Kaneko are both employed by Kikkoman Europe R.&D Laboratory B.V. and by research organizations (WUR and TNO). A possible conflict of interest was prevented by following the WUR-integrity code (URL: https://www.wur.nl/en/About-WUR/Integrity-and-privacy/Scientific-integrity.htm, (accessed on 5 April 2021)). Products from different suppliers were included in the study.

References

1. Keller, K.L. Conceptualizing, measuring, and managing customer-based brand equity. *J. Mark.* **1993**, *57*, 1–22. [CrossRef]
2. Haas, R.; Imami, D.; Miftari, I.; Ymeri, P.; Grunert, K.; Meixner, O. Consumer perception of food quality and safety in western balkan countries: Evidence from albania and kosovo. *Foods* **2021**, *10*, 160. [CrossRef]
3. Dressler, M.; Paunovic, I. The Value of Consistency: Portfolio Labeling Strategies and Impact on Winery Brand Equity. *Sustainability* **2021**, *13*, 1400. [CrossRef]

4. Dressler, M.; Paunovic, I. A Typology of Winery SME Brand Strategies with Implications for Sustainability Communication and Co-Creation. *Sustainability* **2021**, *13*, 805. [CrossRef]

5. Mazzocchi, C.; Orsi, L.; Sali, G. Consumers' attitudes for sustainable mountain cheese. *Sustainability* **2021**, *13*, 1743. [CrossRef]

6. Piqueras Fiszman, B.; Spence, C. Sensory expectations based on product-extrinsic food cues: An interdisciplinary review of the empirical evidence and theoretical accounts. *Food Qual. Prefer.* **2015**, *40*, 165–179. [CrossRef]

7. Skaczkowski, G.; Durkin, S.; Kashima, Y.; Wakefield, M. The effect of packaging, branding and labeling on the experience of unhealthy food and drink: A review. *Appetite* **2016**, *99*, 219–234. [CrossRef]

8. Schifferstein, H.N.; Kole, A.P.; Mojet, J. Asymmetry in the disconfirmation of expectations for natural yogurt. *Appetite* **1999**, *32*, 307–329. [CrossRef] [PubMed]

9. Głuchowski, A.; Czarniecka-Skubina, E.; Kostyra, E.; Wasiak-Zys, G.; Bylinka, K. Sensory Features, Liking and Emotions of Consumers Towards Classical, Molecular and Note by Note Foods. *Foods* **2021**, *10*, 133. [CrossRef]

10. Yeomans, M.R.; Chambers, L.; Blumenthal, H.; Blake, A. The role of expectancy in sensory and hedonic evaluation: The case of smoked salmon ice-cream. *Food Qual. Prefer.* **2008**, *19*, 565–573. [CrossRef]

11. Thomson, D.M.H. Reaching out beyond liking to make new products that people want. In *Consumer Driven Innovation in Food and Personal Care Products*; MacFie, H.J.H., Jaeger, S.R., Eds.; Woodhead: Cambridge, UK, 2010.

12. De Wijk, R.A.; Boesveldt, S. Responses of the autonomic nervous system to flavours. In *Multisensory Flavor Perception: From Fundamental Neuroscience through to the Marketplace*; Piqueras-Fiszman, B., Spence, C., Eds.; Woodhead Publishing Limited: Amsterdam, The Netherlands, 2016; Chapter 11.

13. Viejo, C.G.; Fuentes, S.; Howell, K.; Torrico, D.; Dunshea, F.R. Robotics and computer vision techniques combined with non-invasive consumer biometrics to assess quality traits from beer foamability using machine learning: A potential for artificial intelligence applications. *Food Control* **2018**, *92*, 72–79. [CrossRef]

14. Torrico, D.D.; Fuentes, S.; Viejo, C.G.; Ashman, H.; Gunaratne, N.M.; Gunaratne, T.M.; Dunshea, F.R. Images and chocolate stimuli affect physiological and affective responses of consumers: A cross-cultural study. *Food Qual. Prefer.* **2018**, *65*, 60–71. [CrossRef]

15. Mehta, A.; Sharma, C.; Kanala, M.; Thakur, M.; Harrison, R.; Torrico, D.D. Self-Reported Emotions and Facial Expressions on Consumer Acceptability: A Study Using Energy Drinks. *Foods* **2021**, *10*, 330. [CrossRef]

16. Danner, L.; Joechl, M.; Duerrschmid, K.; Haindl, S. Facial expressions and autonomous nervous system responses elicited by tasting different juices. *Food Res. Int.* **2014**, *64*, 81–90. [CrossRef]

17. Gonzalez, V.C.; Villarreal-Lara, R.; Torrico, D.D.; Rodríguez-Velazco, Y.G.; Escobedo-Avellaneda, Z.; Ramos-Parra, P.A.; Mandal, R.; Pratap, S.A.; Hernández-Brenes, C.; Fuentes, S. Beer and consumer response using biometrics: Associations assessment of beer compounds and elicited emotions. *Foods* **2020**, *9*, 821. [CrossRef] [PubMed]

18. De Wijk, R.A.; He, W.; Mensink, M.M.; Verhoeven, R.; de Graaf, C. ANS responses and facial expressions differentiate between repeated exposures of commercial breakfast drinks. *PLoS ONE* **2014**, *9*, e93823. [CrossRef]

19. Bercík, J.; Paluchová, J.; Neomániová, K. Neurogastronomy as a Tool for Evaluating Emotions and Visual Preferences of Selected Food Served in Different Ways. *Foods* **2021**, *10*, 354. [CrossRef]

20. De Wijk, R.A.; Kaneko, D.; Dijksterhuis, G.B.; van Zoggel, M.; Schiona, I.; Visalli, M.; Zandstra, E.H. Food perception and emotion measured over time in-lab and in-home. *Food Qual. Prefer.* **2019**, *75*, 170–178. [CrossRef]

21. De Wijk, R.A.; Noldus, L.P.J.J. Using implicit rather than explicit measures of emotions. *Food Qual. Prefer.* **2020**, *92*, 104125. Available online: https://edepot.wur.nl/537177 (accessed on 1 May 2021). [CrossRef]

22. Samant, S.S.; Chapko, M.J.; Seo, H.-S. Predicting consumer liking and preference based on emotional responses and sensory perception: A study with basic taste solutions. *Food Res. Int.* **2017**, *100*, 325–334. [CrossRef]

23. Mojet, J.; Durrschmid, K.; Danner, L.; Jochl, M.; Heinio Raija, L.; Holthuysen, N.; Koster, E. Are implicit emotion measurements evoked by food unrelated to liking? *Food Res. Int.* **2015**, *76*, 224–232. [CrossRef]

24. Kaneko, D.; Hogervorst, M.; Toet, A.; van Erp, J.B.F.; Kallen, V.; Brouwer, A.-M. Explicit and Implicit Responses to Tasting Drinks Associated with Different Tasting Experiences. *Sensors* **2019**, *19*, 4397. [CrossRef]

25. Urbano, C.; Mahieu, B.; Thomas, A.; Schlich, P.; Visalli, M. Recording facial mimics and temporal dominance of sensations and emotions. In Proceedings of the Eurosense Conference, Verona, Italy, 2–5 September 2018.

26. Kaneko, D.; Toet, A.; Brouwer, A.-M.S.U.; Kallen, V.; van Erp, J.B.F. EmojiGrid: A 2D pictorial scale for cross-cultural emotion assessment of negatively and positively valenced food. *Food Res. Int.* **2019**, *115*, 541–551. [CrossRef]

27. De Wijk, R.A.; Ushiama, S.; Ummels, M.; Zimmerman, P.; Kaneko, D.; Vingerhoeds, M.H. Effect of branding and familiarity of soy sauces on valence and arousal using facial expressions, physiological measures, emojis, and ratings. *Front. Neuroergonomics* **2021**. [CrossRef]

28. Fenko, A.; Backhaus, B.W.; van Hoof, J.J. The influence of product- and person-related factors on consumer hedonic responses to soy products. *Food Qual. Prefer.* **2015**, *41*, 30–40. [CrossRef]

29. Ushiama, S.; Vingerhoeds, M.H.; Kanemura, M.; Kaneko, D.; de Wijk, R.A. Some insights into the development of food and brand familiarity: The case of soy sauce in the Netherlands. *Food Res. Int.* **2021**, *142*, 110200. Available online: https://edepot.wur.nl/542717 (accessed on 1 May 2021). [CrossRef] [PubMed]

30. Toet, A.; Kaneko, D.; Ushiama, S.; Hoving, S.; de Kruijf, I.; Brouwer, A.-M.; van Erp, J.B.F. Emojigrid: A 2D pictorial scale for the assessment of food elicited emotions. *Front. Psychol.* **2018**, *9*, 2396. [CrossRef]

31. Wei, L.; Tian, Y.; Huang, T.; Wang, Y.; Ebrahimi, T. Automatic webcam-based human heart rate measurements using laplacian eigenmap. In Proceedings of the 11th Asian Conference on Computer Vision ACCV 2012, Daejeon, Korea, 5–9 November 2012.

32. Wang, W.; den Brinker, A.C.; Stuijk, S.; de Haan, G. Algorithmic principles of remote PPG. *IEEE Trans. Biomed. Eng.* **2017**, *64*, 1479–1491. [CrossRef] [PubMed]

33. Kanemura, M.; de Wijk, R.A.; Bittner, M.; den Uyl, T.M.; Zimmerman, P.H.; Theuws, J.J.M. Validation of new methods for exploring consumption behavior: Measuring heart rate through remote PPG and Automated analysis of chewing and biting in FaceReader. In Proceedings of the Pangborn Conference, Edinburgh, UK, 28 July–1 August 2019.

34. Viola, P.; Jones, M. Rapid object detection using a boosted cascade of simple features. In Proceedings of the 2001 IEEE Computer Society Conference on Computer Vision and Pattern Recognition CVPR 2001, Kauai, HI, USA, 18–22 June 2021.

35. Den Uyl, M.J.; Van Kuilenburg, H. The FaceReader: Online Facial Expression Recognition. In Proceedings of the Measuring Behavior 2005, Wageningen, The Netherlands, 30 August–2 September 2008; pp. 589–590.

36. Cootes, T.; Edwards, G.; Taylor, C. Active appearance models. *IEEE Trans. Pattern Anal. Mach. Intell.* **1978**, *23*, 681–685. [CrossRef]

37. Ekman, P.; Friesen, W. *Facial Action Coding System*; Consulting Psychologists Press: Palo Alto, CA, USA, 1978.

38. Bijlstra, G.; Dotsch, R. *FaceReader 4 Emotion Classification Performance on Images from the Radboud Faces Database*; Radboud University: Nijmegen, The Netherlands, 2011.

39. Lewinski, P. Automated facial coding software outperforms people in recognizing neutral faces as neutral from standardized datasets. *Front. Psychol.* **2015**, *6*, 1386. [CrossRef]

40. Scherer, K.R. What are emotions? And how can they be measured? *Soc. Sci. Inf.* **2005**, *44*, 695–729. [CrossRef]

41. Gunaratne, N.M.; Fuentes, S.; Gunaratne, T.M.; Torrico, D.D.; Ashman, H.; Francis, C.; Viejo, C.G.; Dunshea, F.K. Consumer acceptability, eye fixation, and physiological responses: A study of novel and familiar chocolate packaging designs using eye-tracking devices. *Foods* **2019**, *8*, 253. [CrossRef] [PubMed]

42. Desmet, P.M.A.; Schifferstein, H.N.J. Sources of positive and negative emotions in food experience. *Appetite* **2008**, *50*, 290–301. [CrossRef]

43. Hartwell, H.J.; Edwards, J.S.; Brown, L. The relationship between emotions and food consumption (macronutrient) in a foodservice college setting—A preliminary study. *Int. J. Food Sci. Nutr.* **2013**, *64*, 261–268. [CrossRef]

44. Meiselman, H.L. The future in sensory/consumer research: Evolving to a better science. *Food Qual. Prefer.* **2013**, *27*, 208–214. [CrossRef]

45. Poli, S.; Sarlo, M.; Bortoletto, M.; Buodo, G.; Palomba, D. Stimulus-preceding negativity and heart rate changes in anticipation of affective pictures. *Int. J. Psychophysiol.* **2007**, *65*, 32–39. [CrossRef] [PubMed]

46. Symmank, C. Extrinsic and intrinsic food product attributes in consumer and sensory research: Literature review and quantification of the findings. *Manag. Rev. Q.* **2019**, *69*, 39–74. [CrossRef]

Communication

Measuring Implicit Approach–Avoidance Tendencies towards Food Using a Mobile Phone Outside the Lab

Anne-Marie Brouwer [1,*], Jasper J. van Beers [1], Priya Sabu [1], Ivo V. Stuldreher [1], Hilmar G. Zech [2] and Daisuke Kaneko [3]

[1] Department of Human Performance, TNO, 3769 DE Soesterberg, The Netherlands; Jasper.V.Beers@outlook.com (J.J.v.B.); priyasabu1998@hotmail.com (P.S.); ivo.stuldreher@tno.nl (I.V.S.)
[2] Department of Psychology and Neuroimaging Center, Technische Universität Dresden, 01069 Dresden, Germany; hilmar.zech@tu-dresden.de
[3] Kikkoman Europe R&D Laboratory B.V., 6709 PA Wageningen, The Netherlands; d.kaneko@kikkoman.nl
* Correspondence: anne-marie.brouwer@tno.nl

Abstract: Implicit ('unconscious') approach–avoidance tendencies towards stimuli can be measured using the Approach Avoidance Task (AAT). We recently expanded a toolbox for analyzing the raw data of a novel, mobile version of the AAT (mAAT), that asks participants to move their phone towards their face (pull) or away (push) in response to images presented on the phone. We here tested the mAAT reaction time and the mAAT distance in a study with 71 Dutch participants that were recruited online and performed an experiment without coming to the laboratory. The participants used both the mAAT and (explicit) rating scales to respond to photographic images of food. As hypothesized, the rated wanting, rated valence and mAAT reaction time indicated a preference for palatable over unpalatable food, and for Dutch over Asian food. Additionally, as expected, arousal was rated higher for unpalatable than for palatable food, and higher for Dutch than for Asian food. The mAAT distance indicated that the unpalatable food images were moved across larger distances, regardless of the movement direction (pull or push), compared to the palatable food images; and the Dutch food images were moved across larger distances than the Asian food images. We conclude that the mAAT can be used to implicitly probe approach–avoidance motivation for complex images in the food domain. The new measure of mAAT distance may be used as an implicit measure of arousal. The ratings and the mAAT measures do not reflect the exact same information and may complement each other. Implicit measures, such as mAAT variables, are particularly valuable when response biases that can occur when using explicit ratings are expected.

Keywords: food images; consumer; approach–avoidance; Approach–Avoidance Task (AAT); valence; arousal; wanting; implicit measure; self-report; mobile phone

Citation: Brouwer, A.-M.; Beers, J.J.v.; Sabu, P.; Stuldreher, I.V.; Zech, H.G.; Kaneko, D. Measuring Implicit Approach–Avoidance Tendencies towards Food Using a Mobile Phone Outside the Lab. *Foods* **2021**, *10*, 1440. https://doi.org/10.3390/foods10071440

Academic Editor: Damir Dennis Torrico

Received: 16 April 2021
Accepted: 14 June 2021
Published: 22 June 2021

Publisher's Note: MDPI stays neutral with regard to jurisdictional claims in published maps and institutional affiliations.

1. Introduction

Emotional attitudes towards food are considered to be important in predicting consumer behavior [1–5]. It has been shown that, compared to verbal liking preferences, food-evoked emotions have more predictive value in foreseeing whether consumers will like a product or not [1]. Recent literature reviews on the use of implicit ('unconscious') and explicit (self-report) methods to measure food-evoked emotions show the dominance of explicit methods in the field [6,7]. Implicit and explicit measures of food-evoked emotions can convey similar information. For instance, for a range of physiological, behavioral and explicit measures, responses toward tasting a clearly unpalatable drink stand out with respect to responses toward regular drinks [8]. However, on closer examination, all of these measures do reflect different processes. For instance, skin conductance has consistently been found to be positively associated with arousal [9–12], and is influenced by factors unrelated to emotion, such as temperature, whereas explicit reports on arousal reflect 'arousal' as interpreted by the individual, to the extent that he or she is aware of this and chooses to share this information. A difference between explicit and implicit measures, and thus, the added value of implicit measures, is, e.g., expected in cases of social pressure for a certain explicit response, or when explicit responses are affected by cultural bias [13–15].

The tendency to energize behavior towards a positive stimulus or away from a negative stimulus [16] is one of several facets of emotional experience. In the case of food, this approach–avoidance tendency can be estimated by asking individuals their explicit response to whether they want the food. As an implicit measure, Electroencephalogram (EEG) alpha asymmetry has been used [17,18]. Another implicit method, that does not rely on brain signals, is the Approach–Avoidance Task (AAT), first developed by Solarz [19]. He asked participants to pull cards towards themselves, or push them away, and found that cards with positive words were pulled more quickly than cards with negative words, and that cards with negative words were pushed more quickly than cards with positive words. When the original AAT was redesigned to run on personal computers [20,21], this greatly increased the flexibility of the task and facilitated its application across many different research areas. In the redesigned AAT, participants are presented with images on a computer screen and push these 'away' to avoid stimuli or pull them 'near' to approach stimuli by moving a joystick in the direction away or towards themselves, respectively. However, a downside of this version compared to the original, is the ambiguity introduced by the joystick. If one pulls a joystick to oneself, it is ambiguous whether that motion reflects the self (i.e., 'moving myself away from the stimulus', indicating avoidance) or whether the motion reflects the stimulus (i.e., 'moving the stimulus to me', indicating approach). Thus, for a more natural experience, reminiscent of the original test, yet easy to run and quantify, Zech et al. [22] developed a mobile version of the AAT (mAAT), in which images are presented on a smartphone screen that participants have to push away or pull toward themselves. Indeed, it was found that participants were faster when they had to approach positive stimuli (happy faces) or avoid negative stimuli (angry faces), compared to when these instructions were reversed [22]. The mAAT seems a particularly suitable tool to measure approach–avoidance in the domain of food, given that food has a very natural, unambiguous relation to approach and avoidance (bringing food to the mouth, or pushing it away). The fact that the mAAT runs on a mobile phone enables the collection of data outside the lab, which is useful for testing in specific contexts of interest [22] or when coming to the laboratory is impossible or inconvenient for other reasons, such as the COVID-19 pandemic.

As noted by Zech et al. [22], reaction time (RT) may not be the only variable of interest that can be extracted from the mAAT. Participants may not only respond quicker when moving a stimulus in the direction that is congruent to their (approach or avoidance) motivation but may also move these stimuli over a larger distance. The potential advantage of distance over RT is that it may be less sensitive to factors that can affect RT besides approach–avoidance motivation. In cases where complex stimuli are used, such a factor may be the time it takes to recognize a stimulus. We recently improved the usability and analysis of the data generated by the mAAT [23], including calculating the new variable of mAAT distance.

In the food domain, the AAT has been used to investigate healthy eating [24], food craving [25–27] and eating disorders [28]. There are few studies investigating the implicit AAT approach–avoidance tendencies related to food experience. A notable exception is [29]. In this study, a computerized joystick AAT paradigm was used on appealing and disgusting food images, wherein, as expected, the participants exhibited an approach bias towards appealing food and an avoidance bias away from disgusting food.

In the current study, we benchmarked the mAAT and the updated toolbox on photographic images of food. We utilized standardized images [30] for which a very strong difference in approach or avoidance motivation is expected: regular, palatable food (congruent with pull, incongruent with push), and food that was unpalatable because of mold or because it was infested by insects, worms or snails (congruent with push, incongruent with pull). We also used images for which a subtle difference in approach or avoidance motivation is expected: food from the participant's own (in this case, Dutch) culture, and food from another culture (in this case, Asian). Previous studies consistently report that individuals overall prefer familiar food, or food from their own culture [13,15]. Both the mAAT RT and the mAAT distance were examined. The results were related to the explicit measures of approach–avoidance motivation (ratings of wanting) and emotion (valence and arousal) in response to the same set of images.

2. Materials and Methods

2.1. Participants

Participants were recruited through Prolific (www.prolific.co, Prolific, London, UK). In order to participate, participants had to have a Dutch nationality, fall within an age range of 18 to 65 years old and not follow any diet or suffer from any food allergy. See Supplementary File A for the recruitment text. A total of 120 individuals started the procedure. Complete datasets were obtained for 71 participants and were included in the analysis. Thirty of them were female, and their age ranged from 18 to 59, with a median of 30 years old. Their Body Mass Index ranged from 16.5 to 35.5, with a median of 24.5. Most of the participants reported eating Asian food weekly ($n = 33$), followed by monthly ($n = 24$). One participant reported eating Asian food every other day, and the remaining participants ($n = 13$) less than once a month. Participants who completed the experiment received a monetary reward of GBP 5.

2.2. Materials

2.2.1. Stimuli

Food images were taken from the CROCUFID (CROss CUltural Food Images Database; [30]) and represented the following four categories: Asian food, Dutch food, palatable food (i.e., universal food, such as fruits and vegetables) and unpalatable food (i.e., molded food, or food with snails or insects crawling on it). Each category was represented by 20 unique images. Figure 1 shows an example image from each category. The complete set of used images is in Supplementary B.

Figure 1. Example image from each of the four food image categories. Each category was represented by 20 unique images.

2.2.2. Questionnaires

Before the presentation of the food images, the participants filled out a questionnaire that was used to describe the participant sample and to enable the control of possibly relevant factors (such as current feelings of satiation and frequency of eating Asian food—see Supplementary File C). For the same reasons, they also filled out the Food Neophobia Scale [31], consisting of ten questions that the participant rated on a 7-point scale, ranging from 'strongly disagree' to 'strongly agree'. High scoring participants are considered food neophobic, meaning that they are unwilling to try new food, while low scoring participants are enthusiastic about trying new and different food. We used Gorilla (www.Gorilla.sc, Cauldron Science, Cambridge, UK, accessed on 1 June 2020) as the experimental platform to ask the questions and direct the participants through the experiment.

2.2.3. Stimulus Rating Scales

Each food image was rated using two rating scales. We used the EmojiGrid tool [32] to measure explicit food-related valence and arousal. The EmojiGrid is a 2D pictorial scale that separates the valence (x-axis) and arousal (y-axis) axes of emotion. To respond, participants click anywhere on the plane to express their food-related experience. For each trial, we recorded valence and arousal. Participants rated food wanting by using a slider on a VAS (Visual Analogue Scale) running from 'fully disagree' to 'fully agree' in response to the question 'I want this very much'.

2.2.4. mAAT

The mAAT app developed by Zech et al. [22] was set up for our conditions and made available for download from the Google Play Store. Participants installed the app on their personal phone. The app presents images and records the accelerations and rotational rates (if gyroscopic sensors are present) of the phone.

2.3. Experimental Design

Participants performed the experiment in two halves, interleaved with a break during which they were asked to watch a 6-minute movie (One group of participants (n = 38) was asked to watch a movie about the making of Lego bricks; the other group (n = 33) was asked to watch a movie about the making of soy sauce. We suspected that Asian food might be liked better after watching the movie about soy sauce compared to the movie unrelated to food. Since no such effect was observed in any of the variables, in this study, we grouped the data for all analyses.) The experiment halves were identical except for the exact images used, where we divided each of the four sets of 20 images (palatable, unpalatable, Dutch, Asian) into two sets of 10. Which set was presented before the break, and which after the break, was counterbalanced across the participants. Each half consisted of (firstly) the rating task and (secondly) the mAAT.

In the rating task, participants rated the images, presented in random order, using firstly, the EmojiGrid and, secondly, the wanting VAS. The mAAT task consisted of the following two parts: first, the Dutch and Asian food images were presented and, second, the palatable and unpalatable. Before the start of each part, participants were instructed to pull the phone towards them upon presentation of one (randomly determined) type of stimulus (e.g., 'Dutch') and push the phone away upon presentation of the other stimulus type ('Asian'). When all of the images had been shown twice, the opposite instruction was given (i.e., in the example, to pull the phone when an 'Asian' food image was shown and push when 'Dutch' food was presented). Again, all of the images were shown twice. Thus, in the mAAT task, each of the images was presented four times; twice with the instruction to pull and twice with the instruction to push each image. Then, the part with the palatable and unpalatable food images was performed in the same way. After the break, the second half was performed.

2.4. Procedure

Figure 2 depicts the procedure of the complete experiment. Participants read about the experiment in Prolific and signed the informed consent by clicking a checkbox. They could not proceed before giving informed consent. They were then instructed to download the mAAT app from the Appstore on their phone and were redirected to Gorilla on their (desktop or laptop) computer. Participants completed the general questionnaire and Food Neophobia Scale. After that, instructions appeared regarding the rating scales, asking participants to indicate their first impression. Then, the first half of the experiment started. Participants started with rating the food images using the explicit tools. Each image was first presented alongside the EmojiGrid. After clicking the location on the grid that best represented their current emotion towards the presented stimulus using the computer mouse or touchpad, the image was presented again alongside the wanting VAS. After clicking the appropriate location, the next image and scale appeared until all 40 images were rated. Participants were then instructed on the mAAT, including a short movie of the desired type of movements. For each of the four combinations of food types (Asian/Dutch, palatable/unpalatable) and movement instruction (pull 'A'/push 'B' or pull 'B'/push 'A'), participants practiced 5 trials with a dedicated set of (CROCUFID) images from the relevant food categories that were distinct from those used in the experimental trials. Within each trial, participants first saw a fixation cross for 500 ms to guide the eyes to the center of the phone's display. After this, the current trial's image was shown until either the participant responded by moving the phone, or after 2 s had elapsed (this was considered as 'no reaction'). After pushing or pulling the phone, participants completed the response by immediately returning the phone to the initial position. Once the phone had come to rest, the next trial started. After finishing the mAAT, participants were directed to their computer to watch a movie as a break. Then, the second half of the experiment started, which was identical to the first, except for the exact images used. The whole procedure took about 1 h to complete.

2.5. Analysis

For each participant and each stimulus category (palatable, unpalatable, Dutch, Asian), an average score of EmojiGrid valence, EmojiGrid arousal and rated wanting was determined. Wilcoxon signed ranks tests were used to test for significant differences between the palatable and unpalatable, and between the Asian and Dutch food images.

Data from the mAAT app were processed using the expanded mAAT processing toolbox, as described in [23]. The toolbox is freely available for download at https://github.com/Jasper-van-beers/AAT (accessed on 30 November 2020). The mAAT RTs were defined as the time between stimulus onset and onset of the motion of the phone. Motion onset was defined as the moment that the acceleration is greater than the maximum $(0.8, (0.3 \cdot a_{max}))$ ms^{-2}, with a_{max} denoting the maximum measured acceleration. Any RTs < 200 ms were discarded and any RTs > 2000 ms were considered to be 'no reactions'. Data from participants with less than 75% valid trials were considered to be incomplete datasets and were not included in the analyses. The innovative feature of mAAT distance was derived using the magnitude and the duration of the acceleration.

An average RT and an average distance were calculated for each participant, stimulus category and movement direction (pull or push). Repeated measure ANOVAs with stimulus category and movement direction were applied to the mAAT RT and the mAAT distance for the palatable and unpalatable food images, and for the Asian and Dutch food images.

To further explore how implicit mAAT responses relate to other measures that we expect to be associated with the approach and avoidance motivation, we computed an mAAT RT score by subtracting 'mAAT RT pull' from 'mAAT RT push' for each participant and each image category. A high mAAT score would correspond to approach motivation. It was expected to correlate positively with valence and wanting scores, and negatively with food neophobia for Asian food images. Pearson correlations were performed to test for these effects.

Repeated measure ANOVAs were performed using an SPSS 25 (IBM, Armonk, NY, USA). Wilcoxon signed ranks tests and Pearson correlations were performed using a MATLAB R2020a (The MathWorks Inc., Natick, MA, USA). For all statistical tests, we used an alpha level of 0.05.

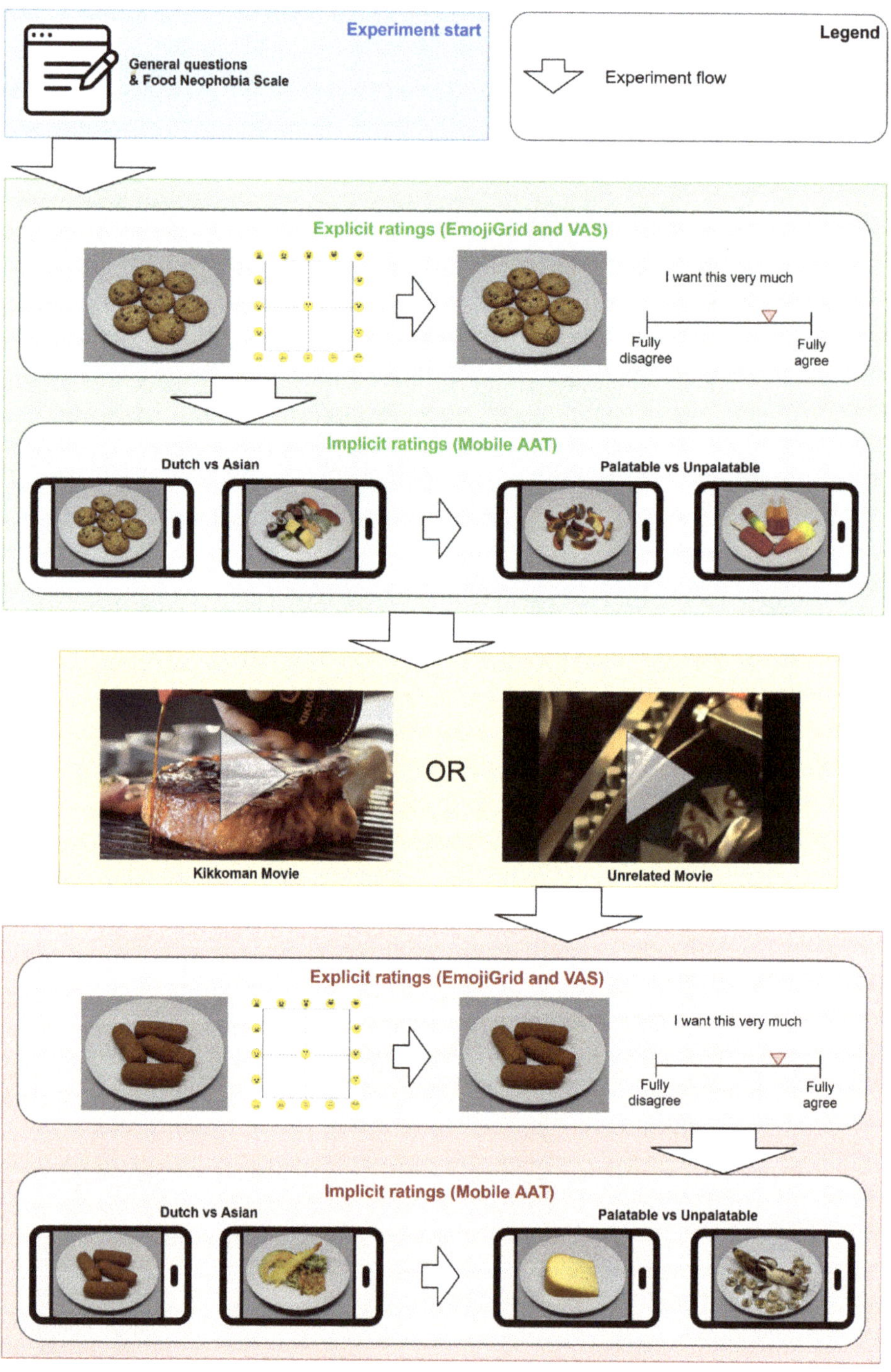

Figure 2. Schematic depiction of the experimental procedure.

3. Results

3.1. Explicit Ratings

Figure 3 shows the explicit ratings of valence (a), arousal (b) and wanting (c), averaged across the participants for each of the four stimulus categories. The Wilcoxon signed ranks tests indicated significant differences between the palatable and unpalatable food images, and between the Asian and Dutch food images, for all three explicit ratings (all p-values < 0.01). The valence and wanting indicated a preference for palatable over unpalatable, and a preference for Dutch over Asian food. The rated arousal was higher for unpalatable than for palatable food, and higher for Dutch than for Asian food.

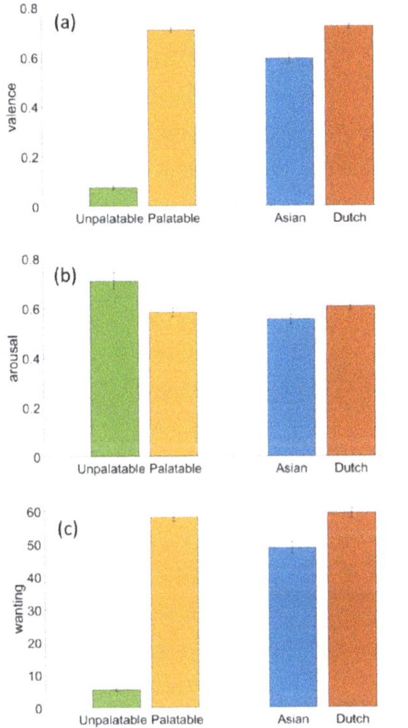

Figure 3. Explicit ratings valence (**a**), arousal (**b**) and wanting (**c**) for each of the four stimulus categories. Error bars indicate standard errors of the mean.

3.2. mAAT Measures

Figure 4 shows the mAAT RT (a) and the mAAT distance (b) averaged across the participants for each of the four stimulus categories and the push–pull direction.

For the mAAT RT, the ANOVA for palatable and unpalatable food showed that, in general, people responded quicker when making a pulling than a pushing movement (main effect of movement direction: $p < 0.001$) and that responses to unpalatable food were quicker (main effect of image type: $p < 0.001$). Importantly, a significant interaction effect between the movement direction and the image type ($p < 0.001$) showed that, as expected, the participants were quicker to push a stimulus congruent with avoidance motivation (i.e., unpalatable food) than a stimulus that was not, relative to pulling. The explicit ratings and the literature led to the expectation that familiar food (Dutch) and unfamiliar food (Asian) result in similar mAAT tendencies as palatable and unpalatable food, respectively. Indeed, the ANOVA for the Asian and Dutch food images showed similar results, with a main effect of the movement direction ($p < 0.001$) and of the image type ($p < 0.001$), as well as an interaction effect ($p = 0.007$), indicating quicker pulling responses than pushing, but especially for the Dutch food images.

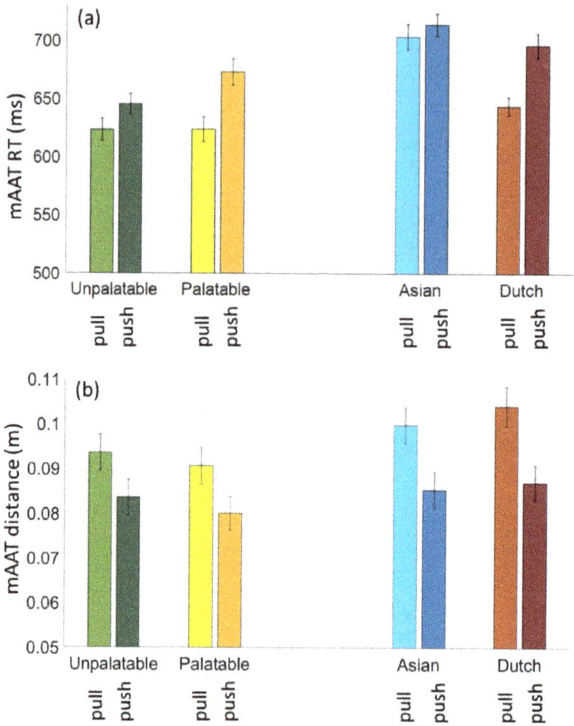

Figure 4. mAAT RT (**a**) and mAAT distance (**b**) for each of the four stimulus categories and each movement direction, pull and push.

For the mAAT distance, the ANOVA for palatable and unpalatable food showed a significant main effect of the movement direction ($p < 0.001$), with shorter distances for pushing than pulling, and a significant effect of the image type ($p < 0.001$), indicating that, overall, the unpalatable food images were moved across larger distances than the palatable images. There was no interaction ($p = 0.79$). The same pattern of results was found for Dutch and Asian food, with a main effect of the movement direction ($p < 0.001$), and a main effect of the image type ($p = 0.001$), where the Asian food images were moved across larger distances compared to the Dutch food images. No interaction effect was present ($p = 0.99$).

3.3. Correlations

The mAAT RT score did not significantly correlate with valence or wanting for any of the four food image categories. It also did not correlate with food neophobia for Asian food images. As a comparison, food neophobia did show a negative correlation with the EmojiGrid valence for Asian food images ($R^2 = 0.32$, $p < 0.001$; Figure 5a), and a similar negative correlation was found between food neophobia and wanting ($R^2 = 0.31$, $p < 0.001$; Figure 5b).

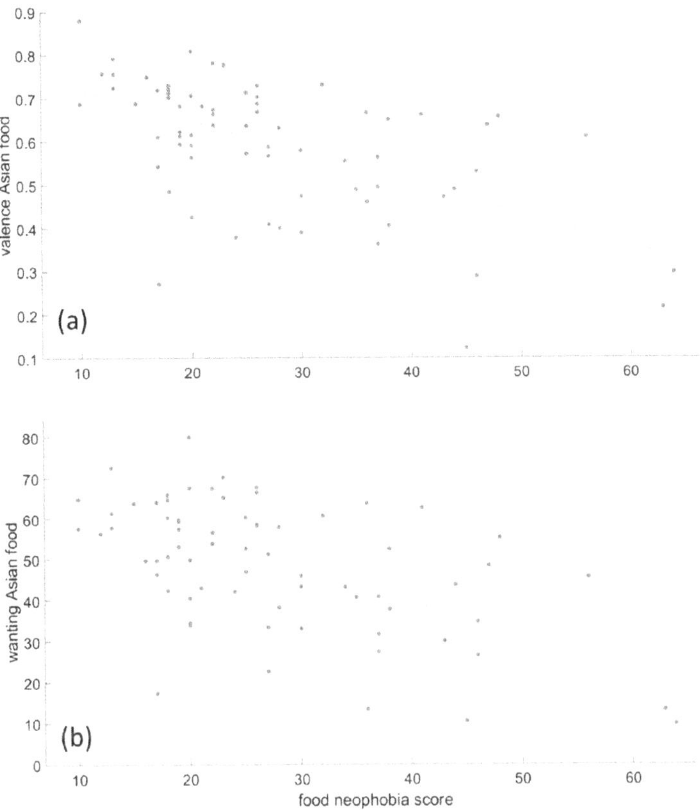

Figure 5. Correlation between food neophobia score and EmojiGrid valence (**a**) and wanting (**b**) for Asian food images. Each data point represents one participant.

Given the effects of the food image categories on the mAAT distance and arousal, we computed an mAAT distance score by averaging the pull and push distance per image category and per participant. These values were correlated to rated arousal for each image category separately, but no significant relations were found.

4. Discussion

The current study showed that approach–avoidance tendencies for food can be reliably measured in participants in the field using a phone, without personal technical help or instructions.

The mAAT RT results showed the expected interaction between an image category and a movement direction, not only for the stimulus categories that were expected to differ strongly in approach–avoidance motivation (palatable and unpalatable food images), but also for more subtly differing food categories (images depicting food from the participant's own or another culture). The explicit ratings of valence, arousal and wanting showed the expected pattern of a strong preference for palatable over unpalatable food, and a preference for their own culture's (Dutch) food over another culture's (Asian) food. While our design did not allow for a direct statistical comparison, as one would expect, the size of the effect in the mAAT RT (i.e., the difference between pull and push), seems to be similar for the palatable and the Dutch food images, whereas the effect seems to be larger for the unpalatable than for the Asian food images. The overall shorter RTs to the palatable and the unpalatable food images compared to the Dutch and the Asian food images may be explained by the difference in the time it takes to identify and categorize the images. It may also be a time order (practice) effect—in each of the two experiment halves, participants responded to the Dutch and Asian food images before the palatable and unpalatable images.

The mAAT distance results showed a different pattern than the mAAT RT results. We had anticipated the mAAT distance to mirror the mAAT RT, i.e., the food images congruent with approach may be pulled both quicker and further towards oneself, and the images congruent with avoidance would be pushed both quicker and further away, where distance may have been relatively unaffected by aspects that are expected to affect RT, such as recognition of the stimulus. However, what we found were larger distances for the unpalatable and the Dutch food images, irrespective of the movement's direction. The unpalatable and the Dutch food images were also judged relatively high in arousal (as found before [15]). Given the specific food images used, depicting molded and infested food, high arousal for the unpalatable images does not come as a surprise. The finding that the Dutch food images were rated higher in arousal than the Asian ones can be understood by the fact that both types of images were generally rated as pleasant, in which case valence and arousal are commonly found to be positively related [33–35]. Since Dutch food is rated high in valence, the high arousal scores are not surprising. The finding that the mAAT distance may be associated with arousal is intriguing and important, since it has been argued that arousal is a crucial determinant in determining (sustained) the attractiveness of products [36,37], but is also hard to capture with explicit questionnaires [35,38]. It would also nicely complement the mAAT RT approach–avoidance motivation, that is more closely related to valence. Future studies need to replicate and further test the possible association between the mAAT distance and arousal.

Given the previous and current results, correlations between rated wanting and mAAT RT, as well as between rated arousal and mAAT distance, may have been expected. However, we did not find such correlations at the participant and stimulus category level. This suggests that these (explicit and implicit) measures reflect different processes. A discrepancy at the condition level may be observed if a discrepancy between the explicit and implicit measures is expected, such as may be the case when there is social pressure to shape explicit responses in a certain way.

A limitation of the study is the loss of participants and data. Twenty-four of the 120 participants that started the procedure quit after performing only a fraction of the experiment. Some of them may not have been able to generate proper mAAT movements. Another twelve participants did not reach the criterion of 75% valid mAAT trials. The number of valid mAAT trials may be increased in the future by setting more strict inclusion criteria for the phones that can be used (e.g., only those containing a linear accelerometer) and by giving participants more precise feedback about inappropriate movements (e.g., rotations rather than pulling and pushing) during the test. In our study, the data of another 13 participants were lost because they did not fill out the rating scales and questionnaires completely or filled out information incompatible with the inclusion criteria. Future online experiments can be made more robust against such omissions by preventing participants from proceeding whenever data is missing or incompatible.

5. Conclusions

In conclusion, the current study showed the sensitivity of the mAAT to measure an approach–avoidance motivation to complex food images, and with the new measure of mAAT distance, possibly arousal, therewith complementing the dominant use of explicit tools in research on food experience. The mAAT more closely maps onto approach–avoidance movement than joystick approaches do. Moreover, the mAAT is a promising tool for evaluating food experience, since it can be used to collect users' implicit tendencies remotely, which can be valuable both from a practical point of view and from a research perspective, when research questions are related to specific times and places that are not compatible with laboratory tests.

Supplementary Materials: The following are available online at https://www.mdpi.com/article/10.3390/foods10071440/s1, Supplementary A: Recruitment, Supplementary B: FoodImages, Supplementary C: Questionnaire.

Author Contributions: Conceptualization, A.-M.B., H.G.Z. and D.K.; methodology, A.-M.B., I.V.S., J.J.v.B. and H.G.Z.; software, J.J.v.B. and H.G.Z.; validation, J.J.v.B.; formal analysis, J.J.v.B., P.S. and D.K.; investigation, J.J.v.B.; resources, H.G.Z.; data curation, J.J.v.B., H.G.Z. and P.S.; writing—original draft preparation, A.-M.B. and J.J.v.B.; writing—review and editing, I.V.S., D.K., P.S. and H.G.Z.; visualization, P.S.; supervision, A.-M.B.; project administration, A.-M.B.; funding acquisition, A.-M.B. and D.K. All authors have read and agreed to the published version of the manuscript.

Funding: This study was funded by Kikkoman Europe R&D Laboratory B.V. Daisuke Kaneko is employed by Kikkoman Europe R.&D Laboratory B.V. and by research organization TNO. Hilmar Zech acknowledges support by the German Research Foundation (Collaborative Research Center TRR 265; Project A02).

Institutional Review Board Statement: The study was conducted according to the guidelines of the Declaration of Helsinki, and approved by the Netherlands Organization for Applied Scientific Research (TNO) Institutional Review Board (protocol code 2020-091, 4 November 2020).

Informed Consent Statement: Informed consent was obtained from all subjects involved in the study.

Conflicts of Interest: The authors declare no conflict of interest.

References

1. Dalenberg, J.R.; Gutjar, S.; Ter Horst, G.J.; de Graaf, K.; Renken, R.J.; Jager, G. Evoked emotions predict food choice. *PLoS ONE* **2014**, *9*, e115388. [CrossRef]
2. Gutjar, S.; De Graaf, C.; Kooijman, V.; De Wijk, R.A.; Nys, A.; Ter Horst, G.J.; Jager, G. The role of emotions in food choice and liking. *Food Res. Int.* **2015**, *76*, 216–223. [CrossRef]
3. Köster, E.P.; Mojet, J. From mood to food and from food to mood: A psychological perspective on the measurement of food-related emotions in consumer research. *Food Res. Int.* **2015**, *76*, 180–191. [CrossRef]
4. Samant, S.S.; Chapko, M.J.; Seo, H.-S. Predicting consumer liking and preference based on emotional responses and sensory perception: A study with basic taste solutions. *Food Res. Int.* **2017**, *100*, 325–334. [CrossRef] [PubMed]
5. Prescott, J. Some considerations in the measurement of emotions in sensory and consumer research. *Food Qual. Prefer.* **2017**, *62*, 360–368. [CrossRef]
6. Kaneko, D.; Toet, A.; Brouwer, A.M.; Kallen, V.; van Erp, J.B. Methods for evaluating emotions evoked by food experiences: A literature review. *Front. Psychol.* **2018**, *9*, 911. [CrossRef]
7. Lagast, S.; Gellynck, X.; Schouteten, J.J.; De Herdt, V.; De Steur, H. Consumers' emotions elicited by food: A sys tematic review of explicit and implicit methods. *Trends Food Sci. Technol.* **2017**, *69*, 172–189. [CrossRef]
8. Kaneko, D.; Hogervorst, M.A.; Toet, A.; van Erp, J.B.F.; Kallen, V.; Brouwer, A.-M. Explicit and implicit responses to tasting drinks associated with different tasting experiences. *Sensors* **2019**, *19*, 4397. [CrossRef]
9. Greenwald, A.G.; Klinger, M.R.; Liu, T.J. Unconscious processing of dichoptically masked words. *Memory Cogn.* **1989**, *17*, 35–47. [CrossRef]
10. Lang, P.J.; Bradley, M.M.; Fitzsimmons, J.R.; Cuthbert, B.N.; Scott, J.D.; Moulder, B.; Nangia, V. Emotional arousal and activation of the visual cortex: An fMRI analysis. *Psychophysiology* **1998**, *35*, 199–210. [CrossRef]
11. Lang, P.J.; Greenwald, M.K.; Bradley, M.M.; Hamm, A.O. Looking at pictures: Affective, facial, visceral, and behavioral reactions. *Psychophysiology* **1993**, *30*, 261–273. [CrossRef]
12. Brouwer, A.-M.; Van Den Broek, T.; Hogervorst, M.; Kaneko, D.; Toet, A.; Kallen, V.; van Erp, J.B.F. Estimating affective taste experience using combined implicit behavioral and neurophysiological measures. *IEEE Trans. Affect. Comput.* **2020**. [CrossRef]
13. Torrico, D.D.; Fuentes, S.; Viejo, C.G.; Ashman, H.; Dunshea, F.R. Cross-cultural effects of food product familiarity on sensory acceptability and non-invasive physiological responses of consumers. *Food Res. Int.* **2019**, *115*, 439–450. [CrossRef]
14. Torrico, D.D.; Fuentes, S.; Viejo, C.G.; Ashman, H.; Gunaratne, N.M.; Gunaratne, T.M.; Dunshea, F.R. Images and chocolate stimuli affect physiological and affective responses of consumers: A cross-cultural study. *Food Qual. Prefer.* **2018**, *65*, 60–71. [CrossRef]
15. Kaneko, D.; Stuldreher, I.; Reuten, A.; Toet, A.; van Erp, J.B.; Brouwer, A.-M. Comparing explicit and implicit measures for assessing cross-cultural food experience. *Front. Neuroergon.* **2021**, *2*, 646280. [CrossRef]
16. Elliot, A.J. The Hierarchical Model of Approach-Avoidance Motivation. *Motiv. Emot.* **2006**, *30*, 111–116. [CrossRef]
17. Harmon-Jones, E.; Gable, P.A.; Peterson, C.K. The role of asymmetric frontal cortical activity in emotion-related phenomena: A review and update. *Biol. Psychol.* **2010**, *84*, 451–462. [CrossRef]
18. Brouwer, A.; Hogervorst, M.; Grootjen, M.; Van Erp, J.; Zandstra, E. Neurophysiological responses during cooking food associated with different emotions. *Food Qual. Prefer.* **2017**, *62*, 307–316. [CrossRef]

19. Solarz, A.K. Latency of instrumental responses as a function of compatibility with the meaning of eliciting verbal signs. *J. Exp. Psychol.* **1960**, *59*, 239. [CrossRef] [PubMed]

20. Chen, M.; Bargh, J.A. Consequences of automatic evaluation: Immediate behavioral predispositions to approach or avoid the stimulus. *Personal. Soc. Psychol. Bull.* **1999**, *25*, 215–224. [CrossRef]

21. Rinck, M.; Becker, E.S. Approach and avoidance in fear of spiders. *J. Behav. Ther. Exp. Psychiatry* **2007**, *38*, 105–120. [CrossRef] [PubMed]

22. Zech, H.G.; Rotteveel, M.; van Dijk, W.W.; van Dillen, L.F. A mobile approach-avoidance task. *Behav. Res. Methods* **2020**, *52*, 2085–2097. [CrossRef] [PubMed]

23. van Beers, J.; Kaneko, D.; Stuldreher, I.; Zech, H.G.; Brouwer, A.M. An accessible tool to measure implicit approach-avoidance tendencies towards food outside the lab. In Proceedings of the MHFI4—4th Workshop on Multisensory Approaches to Human-Food Interaction, Held in Conjunction with the 22th ACM International Conference on Multimodal Interaction, Utrecht, The Netherlands, 25–29 October 2020.

24. Becker, D.; Jostmann, N.B.; Wiers, R.W.; Holland, R.W. Approach avoidance training in the eating domain: Testing the effectiveness across three single session studies. *Appetite* **2015**, *85*, 58–65. [CrossRef]

25. Kemps, E.; Tiggemann, M.; Martin, R.; Elliott, M. Implicit approach–avoidance associations for craved food cues. *J. Exp. Psychol. Appl.* **2013**, *19*, 30. [CrossRef] [PubMed]

26. Brockmeyer, T.; Hahn, C.; Reetz, C.; Schmidt, U.; Friederich, H.-C. Approach bias and cue reactivity towards food in people with high versus low levels of food craving. *Appetite* **2015**, *95*, 197–202. [CrossRef]

27. Meule, A.; Lender, A.; Richard, A.; Dinic, R.; Blechert, J. Approach–avoidance tendencies towards food: Measurement on a touchscreen and the role of attention and food craving. *Appetite* **2019**, *137*, 145–151. [CrossRef]

28. Paslakis, G.; Kühn, S.; Schaubschläger, A.; Schieber, K.; Röder, K.; Rauh, E.; Erim, Y. Explicit and implicit approach vs. avoidance tendencies towards high vs. low calorie food cues in patients with anorexia nervosa and healthy controls. *Appetite* **2016**, *107*, 171–179. [CrossRef]

29. Piqueras-Fiszman, B.; Kraus, A.A.; Charles Spence, C. "yummy" versus "yucky"! explicit and implicit approach–avoidance motivations towards appealing and disgusting foods. *Appetite* **2014**, *78*, 193–202. [CrossRef]

30. Toet, A.; Kaneko, D.; de Kruijf, I.; Ushiama, S.; van Schaik, M.G.; Brouwer, A.M.; Kallen, V.; Van Erp, J.B. CROCUFID: A cross-cultural food image database for research on food elicited affective responses. *Front. Psychol.* **2019**, *10*, 58. [CrossRef]

31. Pliner, P.; Hobden, K. Development of a scale to measure the trait of food neophobia in humans. *Appetite* **1992**, *19*, 105–120. [CrossRef]

32. Kaneko, D.; Toet, A.; Ushiama, S.; Brouwer, A.-M.; Kallen, V.; van Erp, J.B.F. Emojigrid: A 2d pictorial scale for cross-cultural emotion assessment of negatively and positively valenced food. *Food Res. Int.* **2019**, *115*, 541–551. [CrossRef]

33. Marchewka, A.; Zurawski, Ł.; Jednoróg, K.; Grabowska, A. The Nencki Affective Picture System (NAPS): Introduction to a novel, standardized, wide-range, high-quality, realistic picture database. *Behav. Res. Methods* **2014**, *46*, 596–610. [CrossRef]

34. Riegel, M.; Moslehi, A.; Michałowski, J.M.; Zurawski, Ł.; Horvat, M.; Wypych, M.; Jednoróg, K.; Marchewka, A. Nencki Affective Picture System: Cross-cultural study in Europe and Iran. *Front. Psychol.* **2017**, *8*, 274. [CrossRef] [PubMed]

35. Toet, A.; Kaneko, D.; Ushiama, S.; Hoving, S.; de Kruijf, I.; Brouwer, A.M.; Kallen, V.; Van Erp, J.B.F. EmojiGrid: A 2D Pictorial Scale for the Assessment of Food Elicited Emotions. *Front. Psychol.* **2018**, *9*, 2396. [CrossRef] [PubMed]

36. Köster, E.P.; Mojet, J. Theories of food choice develop ment. In *Understanding Con Sumers of Food Products*; Frewer, L.J., Trijp, J.C.M., Eds.; Woodhead Publishing: Cambridge, UK, 2006; pp. 93–124.

37. Köster, E.P.; Mojet, J. Boredom and the reasons why some new products fail. In *Consumer-Led Food Product Development*; MacFie, H., Ed.; Woodhead Publishing Limited: Cambridge, UK, 2007; pp. 262–280.

38. Ribeiro, R.L.; Pompéia, S.; Bueno, O.F.A. Comparison of Brazilian and American norms for the international affective picture system (IAPS). *Rev. Braz. Psychiatr.* **2005**, *27*, 208–215. [CrossRef] [PubMed]

foods

Article

Consumer Acceptance of a Ready-to-Eat Meal during Storage as Evaluated with a Home-Use Test

Maria Laura Montero [1,2], Dolores Garrido [3,4], R. Karina Gallardo [3,5], Juming Tang [6] and Carolyn F. Ross [1,*]

1   School of Food Science, Washington State University, Pullman, WA 99164, USA; maria.montero@wsu.edu
2   National Center for Food Science and Technology (CITA), University of Costa Rica, San José 11501-2060, Costa Rica
3   School of Economic Sciences, Washington State University, Pullman, WA 99164, USA; garridom@union.edu (D.G.); karina_gallardo@wsu.edu (R.K.G.)
4   Department of Economics, Union College, Schenectady, NY 12305, USA
5   School of Economics, Puyallup Research and Extension Center, Washington State University, Puyallup, WA 98371, USA
6   Department of Biological Systems Engineering, Washington State University, Pullman, WA 99164, USA; jtang@wsu.edu
*   Correspondence: cfross@wsu.edu; Tel.: +1-(509)-335-2438

Abstract: A home-use test (HUT) is one method that provides a measure of ecological validity as the product is consumed in home under common daily use circumstances. One product that benefits from being evaluated in-home are ready-to-eat (RTE) meals. This study determined consumer acceptance of microwave-thermally-pasteurized jambalaya, a multi-meat and vegetable dish from American Cajun cuisine, and a control (cooked frozen jambalaya) through an on-line home-use test (HUT) over a 12-week storage period. Paralleling the HUT, an online auction determined consumers' willingness to pay. The study also explored how the social environment may impact the liking of the meals when a partner of the participants joined the sensory evaluation of the meals. Consumers (n = 50) evaluated microwave-processed jambalaya stored at 2 °C and a control (cooked frozen jambalaya stored at −31 °C) after 2, 8 and 12 weeks of storage. Consumer liking of different sensory attributes was measured. Participants could choose to share the meals with a partner as a way to enhance ecological validity. The responses from 21 partners to the sensory-related questions were collected. After the sensory evaluation, the participants bid on the meal they had just sampled. Results showed that processing method (microwave vs. control) did not significantly influence the measured sensory attributes. Only flavor liking decreased over storage time ($p < 0.05$). The inclusion of partners significantly increased ($p = 0.04$) the liking of the appearance of the meals. The mean values of the bids for the meals ranged from \$3.33–3.74, matching prices of commercially available jambalaya meals. This study found suggests that the shelf-life of microwave-processed meals could be extended up to 12 weeks without changing its overall liking. The study also shows the importance of exploring HUT methodology for the evaluation of consumers' acceptance of microwave-processed jambalaya and how including a partner could contribute to enhance ecological validity.

Keywords: home-use test; ecological validity; jambalaya; online auction

Citation: Montero, M.L.; Garrido, D.; Gallardo, R.K.; Tang, J.; Ross, C.F. Consumer Acceptance of a Ready-to-Eat Meal during Storage as Evaluated with a Home-Use Test. *Foods* **2021**, *10*, 1623. https://doi.org/10.3390/foods10071623

Academic Editor: Damir Torrico

Received: 15 June 2021
Accepted: 11 July 2021
Published: 13 July 2021

Publisher's Note: MDPI stays neutral with regard to jurisdictional claims in published maps and institutional affiliations.

1. Introduction

Sensory evaluations are commonly made in confined spaces in sensory laboratories. However, people do not habitually consume their meals in sensory booths, individually partitioned in a room full of strangers, focused on a ballot to assess their food. Rather, they eat at home, in a restaurant or café; sometimes alone, but frequently with family, friends and colleagues [1].

Koster (2003) [2] has shown that testing in a sensory lab regular setup removes the natural consumption setting in which a food is consumed, and thus makes it difficult to elicit accurate data from consumers. Ecological validity is the degree to which a test predicts behaviors in real-world settings or the extent to which the context of the evaluation matches the user's real context [3]. Test designs that parallel real-life situations produce findings that can be generalized to real life outcomes [4]. To enhance ecological validity, the site of the research can be shifted from the standard sensory booth to a setting closer to a more authentic environment, one that closely approximates the condition in which consumers actually purchase or consume the food products being tested [1,4,5].

Because it remains challenging to evaluate food products in a controlled manner within the consumer's natural habitat, in recent years researchers have begun to explore novel solutions to provide more authentic environments [1]. The home-use-test (HUT) is one such methodology. In a HUT, consumers prepare, consume, and evaluate samples in their homes, usually for a period of several days [6,7]. HUTs offer tremendous advantages in terms of validity of the data generated, this type of test is less controlled and allows for the evaluation of product attributes under conditions that relate more closely to real-life usage, thereby increasing the validity of data obtained [6,8]. The opinions of other family members or partners can also enter the picture, as they do in everyday use of purchased products.

Various factors, such as social interaction, physical environment and the serving of the product may influence the liking of different food products [9]. Social interaction may play a key role in consumer behavior [7]. Social interaction may explain the reason why different hedonic results could be observed between standardized situation tests (SST), like laboratory test or central location tests (CLT) and HUTs. In SST, the consumption of the product been tested is always individual whereas in HUT it can be social [7]. Clendenen, Herman, and Polivy, 1994 [10] reported that subjects eat more when eating occurred in groups of several people than when they eat alone especially when meal companions are relatives or friends.

The different locations, even when consumers evaluate food under natural conditions, can also impact the acceptance of a food product depending on the circumstances [7]. Accordingly, de Graaf et al. (2005) [11] reported that the predictive ability of laboratory ratings depends on the type of food been evaluated. Lab ratings are more relevant for snacks than for served dishes. A key factor in the explanation of differences in hedonic results between SST and HUT is the way the product is usually eaten [7]. For products that are strongly related to specific contexts and serving size, a HUT might be more useful to determine consumers' acceptance [6]. Multiple studies have compared CLT versus HUTs in the evaluation of consumers' acceptance of products such as ready-to-mix protein beverages [6]; cod products [12]; ready-to-heat meals [13]; as well as salted cheese crackers and sparkling water [7].

Auctions offer another ecologically valid method of assessing consumer acceptance of food. Although individual dietary choices are primarily influenced by such considerations as taste, convenience and nutritional value of foods [14] cost has to be considered as a factor in the development of the product. In new product development, it is important to evaluate consumers' willingness to pay for the product. In laboratory settings or field experiments, such as HUTs, auctions have been intensively employed to elicit willingness to pay. In auctions, products, services or rights are bought and sold through a formal bidding process [15]. An auction is useful to gain knowledge of consumers' evaluations of a product or brand; thus, auctions can be used to reveal consumers' valuations to facilitate future pricing decisions [16].

One product that is appropriate for assessment by a HUT and an auction is a RTE meal. Ready-to-eat (RTE) meals are products that are pre-cooked, packaged, and ready for consumption without additional preparation and cooking beyond simple heating [17]. Consumption of RTE meals in the United States has been influenced by the fact that over the past four decades, demand has grown for foods that save households time in meal preparation and cleanup (i.e., "convenience foods"). This type of meal fits very well with the needs of consumers who are looking for convenience in food products [18]. However, there is also a need to develop more nutritious, safe, autochthonous, and quality-enhanced RTE meals.

Within this context, microwave pasteurization offers opportunities for the food industry to produce high quality, safe, frozen and chilled RTE meals [19]. The main advantage of a microwave-assisted pasteurization system (MAPS) over traditional thermal processing systems is reduced processing time; the generation of volumetric heat in this system makes it possible to increase the heat transfer rate and reduce the total heating time by three to five times [20,21]. Thus, MAPS is particularly suitable for pasteurization of pre-packaged, heat-sensitive, multi-component meals that are highly viscous, semisolid, or solid [19]. Montero, Sablani, Tang and Ross (2020) [22] investigated the potential of MAPS to extend the shelf life of RTE fried rice. The authors found that MAPS processing was able to extend the shelf life of a chilled fried rice meal up to 6 weeks when stored at 7 °C, demonstrating the potential of this technology for the RTE industry. Barnett, Sablani, Tang and Ross (2019) [23] evaluated the shelf life of sterilized microwave-processed chicken meals and consumer liking of the meals and found that the overall liking did not vary due to the effect of storage time.

The use of a sensory methodology such as a HUT is highly suitable to evaluate consumers' liking of RTE meals. Since there is an extra step before the consumption of the meal (e.g., heating via microwave), and it is tested at home this adds a more realistic context to the sensory experience and expectations of consumers because it resembles the way these meals are usually eaten.

Therefore, this study determined consumer acceptance of MAPS-processed jambalaya and a control (cooked and frozen jambalaya) through an on-line HUT over a 12-week storage period. Jambalaya was chosen because it is a multicomponent ready-to-eat meal with three different types of protein ingredients (sausage, chicken and shrimp) and a vegetable-based sauce, making it a complete meal and suitable for microwave processing. Jambalaya, a regional dish of the American South, is a type of RTE meal that is increasingly available nationwide to consumers who are interested in exploring regional and global cuisines [24,25]. Paralleling the HUT, an online auction determined consumers' willingness to pay. Another goal of the study was to determine the degree to which a manipulation in the social environment of the HUT impacted the level of perceived acceptability on the part of the participants

The study had two hypotheses: (1) the acceptance/liking of different sensory characteristics of MAPS-processed RTE jambalaya would not change significantly during storage as compared to a control (cooked and frozen jambalaya) over a 12-week storage period; (2) ecologically valid measures of consumer acceptance (a modified HUT and an auction) would impact the degree of acceptance of the RTE meals.

2. Materials and Methods

2.1. Preparatory Steps

2.1.1. Jambalaya RTE Meals Preparation

The formulation and ingredients shown in Table 1 were used in the production of the jambalaya:

The jambalaya was manufactured in a sanitary food preparation room at the School of Food Science facilities (Pullman, WA, USA) according to the formulation presented in Table 1. Over a period of three days, 120 trays were assembled and sealed per day for a total of 360 jambalaya trays. The total time for a batch of 120 trays to be cooked, assembled, stored, and MAPS-processed was two days. The workload was staggered in batches of 120 trays over a four-day period as shown in Table 2.

Table 1. Jambalaya formulation and ingredient specifications.

Ingredient	Percentage (%)	Source/Brand
Vegetables		
Crushed tomatoes	34.12	Signature Select
Pre-chopped yellow onion	8.77	Safeway brand
Pre chopped celery ribs	4.09	Safeway brand
Pre chopped *pasilla* pepper	4.02	Safeway brand
Ready-to-use minced garlic	0.66	Spice world
Meat		
Chicken breast	13.32	Safeway brand
Raw, small, shell and tail-on shrimp	9.97	Waterfront Bistro
Andouille pork smoked sausages	9.03	Johnsonville
Spices		
Worcestershire sauce	0.64	Lea and Perrin
Old Bay seasoning	0.11	McCormick
Cajun seasoning	0.10	IGA brand
Salt	0.05	Morton
Coarse ground black pepper	0.03	McCormick
Other		
Chicken broth	15.09	Swanson
Total	100.00	

Table 2. Preparation and processing schedule of the three batches of jambalaya used in the study.

Steps in the Process	Day 1	Day 2	Day 3	Day 4
Cooking/assembling	B1 [1]	B2 [2]	B3 [3]	
MAPS processing		B1	B2	B3
Freezing of the controls		B1	B2	B3

[1] B1: Batch 1; [2] B2: Batch 2; [3] B3: Batch 3.

All the ingredients were weighed and prepared the day before the preparation of the jambalaya. The daily cooking steps for the chicken, sausage, shrimp and sauce are described here.

Chicken: 15 mL of olive oil was added to a deep sauté pan over medium-high heat (level 6). The chicken was added and cooked for 5 min on one side until golden brown. Each piece of chicken was then turned over and cooked for another 5 min until fully cooked (to an internal temperature of 74 °C). The cooked chicken was transferred to a bowl and set aside for 5 min. Then the meat was manually pulled into pieces of approximately 2.54 cm (1 inch). The pulled chicken pieces were transferred to a disposable aluminum foil pan with a lid and then stored at 4 °C until the trays were assembled.

Sausage: These were unpacked, and the edges were cut and discarded. They were sliced into 0.6 cm rounds. Approximately 14–16 slices were obtained from each sausage. A total of 15 mL of olive oil was heated in a pan over medium-high heat (level 6/10). Then the sausage slices were added. They were seared for 2 min on one side. The pan was then removed from the burner and the sausages were turned to the uncooked side. The uncooked side was seared for 30 s. Each batch of cooked sausages was then transferred into a disposable aluminum pan with a lid and stored at 4 °C until the trays were assembled.

Shrimp: The shrimp were thawed by placing 48–54 shrimp on trays the day before the jambalaya preparation, so they could defrost for at least 16 h under refrigerated conditions (4 ± 1 °C). After being thawed, the shells were removed but not the tails. A total of 15 mL of olive oil was heated in a pan over low-medium heat (level 4/10). Then the shrimp were added. The shrimp were seared on one side for 1 min. The pan was then removed from the burner and the shrimp were turned to the uncooked side. Then the shrimp were seared for 30 s on the uncooked side. Each batch of cooked shrimp was then transferred into a disposable aluminum pan with a lid and stored at 4 °C until the trays were assembled.

Sauce: 15 mL of olive oil was added to a large pot over medium-high heat (level 6/10). Then the pre-chopped onion, celery, and *pasilla* pepper were added. The ingredients were cooked until they caramelized (approximately 6 min). Next the chicken broth, Old Bay and Cajun seasoning, tomatoes and Worcestershire sauce were added. The heat was increased and brought to a boil (approximately 15 min). Then the heat was reduced to simmer (low heat level 2–3) for 5–6 min. Each batch of cooked sauce was then transferred to a disposable aluminum pan with a lid and stored at 4 °C until the trays were assembled.

Meal Assembly: After all the meat ingredients (chicken, sausages, and shrimp) and the sauce were prepared, 250 ± g EVOH trays (Silgan PFC, dimensions: 15.5 × 11 × 3 cm) were assembled as described here. The assembled product in each tray consisted of 30 ± 0.5 g of sausages (6 units); 40 ± 0.5 g of shrimp (6 units); 40 ± 0.5 g of pulled chicken; and 140 ± 0.5 g of sauce. Once the trays were assembled, they were sealed with film lids with the same composition reported by Barnett et al. (2019) [23], under the following conditions: 200 °C for 4 s under a 65 mbar vacuum with a 400 mbar nitrogen flush. The sealed trays were then stored at 4 °C.

2.1.2. MAPS and Frozen Meal Processing

MAPS Processing/Freezing: On the day following production, 60 trays of the daily production of 120 were processed through MAPS and the other half were frozen (−35 °C) and used as a control. In total, 180 were frozen as controls and 180 trays were pasteurized in a pilot-scale MAPS in the Food Processing Pilot Plant at Washington State University (WSU), Pullman, WA. A detailed description of MAPS can be found in Tang et al. (2018) [19]. The specific processing conditions used to produce jambalaya in the MAPS are described in the methods section of Montero et al., (2020) [22]. At the time the study was conducted, the MAPS could process 16 trays in one run; thus, there were 12 total runs. After being MAPS-processed, the 180 trays were stored at 2.0 ± 0.5 °C. A Temperature Data Logger RC-5+ (Elitech, CA, USA) was used to track the storage temperature during the whole study.

A total of 180 trays were used as control samples (frozen and stored at −31 °C). The control samples were sealed under conditions identical to those of the MAPS-samples. During the freezing step, the sample trays were placed on boards across the top shelves in a freezer 1 m in front of the evaporator with an air velocity of 1.6 m/s and stored at −31 °C. The storage conditions for the control samples were selected to ensure minimal product changes over the length of the study.

Trays of each type (MAPS and control) were randomly selected and analyzed for microbial, sensory, and chemical properties at 2, 8, and 12 weeks of storage.

Microbial/Safety Testing: At weeks 2, 8 and 12 microbial analyses were performed. MAPS-processed jambalaya and control trays were randomly selected and sent to Microchem Laboratories (Seattle, WA, USA). The jambalaya samples were screened for the following pathogens as a way to assure their safety before human consumption, *Bacillus cereus* (Local Instruction); *Salmonella*; *Listeria monocytogenes*; and *E. coli* O157:H7 (AOAC 050501). For the analyses of pathogens, a 25 g sample was tested. The following analyses from AOAC International Official Methods of Analysis were used to detect signs of spoilage: aerobic plate count; yeasts and molds; and total coliforms. The results from the microbial testing are presented in Table 3.

Based on the microbial testing results the jambalaya meals were safe for consumption at each of the evaluated time points.

The jambalaya meals were evaluated in two separate sensory evaluations by two different groups of participants, a home-use test and a semi-trained panel evaluation.

Table 3. Pathogens and spoilage-related microbial analyses conducted on the MAPS-processed jambalaya and the control during 12 weeks of storage at 2 °C and −31 °C, respectively.

Storage Time (Weeks)	Treatment	Microorganism Tested						
		B. cereus (CFU/g)	*Salmonella* (in 25 g)	*L. monocytogenes* (in 25 g)	*E. coli* O157:H7 (in 25 g)	Aerobic Count (CFU/g)	Yeast and Molds (CFU/g)	Total Coliforms (CFU/g)
2	MAPS [1]	<10	Negative	Negative	Negative	40	<10	<10
	Control	<10	Negative	Negative	Negative	10	<10	<10
8	MAPS	<10	Negative	Negative	Negative	10	<10	10
	Control	<10	Negative	Negative	Negative	21	<10	<10
12	MAPS	<10	Negative	Negative	Negative	<10	<10	<10
	Control	<10	Negative	Negative	Negative	86	<10	<10

[1] Microwave-thermally processed jambalaya is represented as MAPS and cooked frozen jambalaya is represented as Control.

2.1.3. Participant Recruitment and Orientation

The study protocol described here received the approval of the WSU Institutional Review Board for conducting tests with human subjects, under the title Consumer Preferences of Jambalaya IRB #16994.

Participant Recruitment and Selection: 50 participants with previous experience in sensory evaluation (18 male, 32 female, ages 21 to 78 years, mean age = 40 years) were recruited through the WSU Sensory Evaluation Listserv. Most of the participants were students, staff, or retirees of WSU and community members living in the Pullman (WA) and Moscow (ID) region.

The participants were recruited based on the following three criteria: expressed liking for and frequency of consumption of RTEs (at least twice a month); not presenting allergies to the jambalaya ingredients; and being available and committed to doing the sensory testing at the three defined time points.

Orientation and Procedures for the HUT: The jambalaya samples were tested in a home-use test. A 30 min orientation session was conducted on the same day of the first sensory evaluation time point. The objectives of the session were to explain to the participants the general aim of the study; to provide instructions on how they should manage the two jambalaya samples prior to and during consumption (e.g., heating instructions); to explain how the sensory evaluation was conducted online; and to explain how to participate in the online auction (Figure 1).

Because the jambalaya samples provided a full serving so that two adults could portion out and evaluate the same serving, participants who were able to have one other person evaluate the samples with them (i.e., partner, husband, wife, friend, roommate) were encouraged to do so. The requirements for a partner to participate were to be over 18 years old and not present allergies to the jambalaya ingredients. All participants' partners signed a consent form in accordance with IRB #16994. A total of 21 partners joined the study (11 male, 10 female, ages 26 to 85 mean age = 44 years). Their answers to the sensory evaluation were collected with a paper-based questionnaire. Partners did not participate in the on-line auction.

As shown in Figure 1, the participants picked up the two jambalaya samples from the Food Science and Human Nutrition Building on the specified evaluation day. At each time point, the samples were provided to the participants in a small cooler that contained the two jambalaya samples packed inside a plastic bag with a sticker that indicated the heating instructions and the order in which the samples should be tested. Each jambalaya meal was assigned a three-digit code so the participants could easily identify each meal. The serving order was randomized across participants. In the heating instructions, the participants were asked to first puncture the tray's lid on each corner using a knife; then to microwave the meal on Power 9 for 3 min; let the sample rest for 1 min inside the microwave; afterwards to take the tray out of the microwave and to carefully stir the content with a spoon; finally, to transfer the content to a white container so they could easily conduct the sensory evaluation. The participants were indicated to evaluate the jambalaya meals during dinner time, between 5:00 and 9:00 p.m.

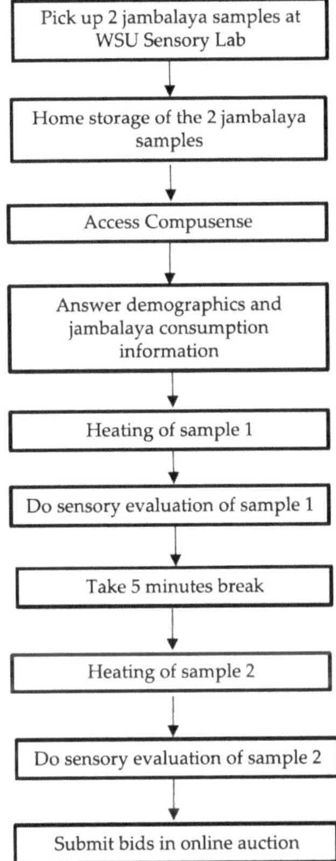

Figure 1. Instructions provided to the panelists in the home-use test evaluation of jambalaya.

The control trays had been thawed in water at room temperature for 8 h before being distributed to the participants, so both the MAPS sample and the control looked the same. The cooler also contained an ice pack (Freez Pak™ Mini, Lifoam, MD, USA) to keep the jambalaya samples at a cool temperature and 2 units of unsalted crackers (Nabisco, NJ, USA) were provided to serve as palate cleansers.

2.2. Evaluation Procedures

2.2.1. HUT Evaluation

HUT Scales: Participants in the home-use test used a total of four different scales to evaluate the entrees described here. Question design and data acquisition were accomplished with Compusense® Cloud (Guelph, ON, Canada) software.

A 7-point hedonic scale [13] was used to test the liking/acceptance of different sensory modalities: the overall liking; aroma; overall flavor; texture acceptance of the shrimp, chicken and sausage; and the overall liking.

A 5-point just-about-right (JAR) scale was used to test the spiciness and texture perception of each of the three meat components (shrimp, chicken and sausage).

A 3-point JAR scale about perception of the size of the jambalaya meal was asked at the end of the study. The scale ranged from 1 (=less than I would like) to 3 (=more than I would like). The participants were asked about their perception of the unit/tray size (250 g = 1 serving); the quantity of sauce; the quantity of vegetables; the size of the vegetables; the quantity of each of the meats, shrimp, chicken, and sausage; the level of saltiness; and their preference for tails off the shrimp.

Participants were also asked (open question) to describe the experience participating in the HUT. Comments were collected, revised and categorized into seven groups. The categories were validated by the agreement between two researchers of the study. The categories are the following: enjoyed experience with partner, HUT vs. in-lab evaluation, time flexibility, fun/positive experience, liking of the meals and willingness to pay for the meals.

Willingness to Pay Evaluation: In collaboration with the School of Economics, a complementary study, an online auction was conducted to measure product satisfaction by the willingness of participants to pay for the jambalaya samples. At each of the three evaluation time points (Weeks 2, 8 and 12), after the sensory evaluation component, the participants were asked to submit their bids (i.e., their willingness to pay) for a unit (equivalent to 9 oz-250 g) of each of the jambalaya sample tested. Compensation for doing the sensory evaluations as well as the online auction at each of the three evaluation points totaled $90.00 in cash mailed to the participants. Partners were not included for this component of the study.

The online auction followed a second price auction protocol. The protocol and the benefits of using this type of action to determine the willingness to pay are described by Lusk and Shogren (2007) [26].

The protocol followed in the present study is reported by Garrido et al. (2021) [24]. To determine the winner of the auction, the first step was to randomly select one of the jambalaya samples (control or MAPS). The winner of the auction was the participant who placed the highest bid for the selected sample. The winner received one meal unit of this meal, and in exchange, they had to pay the market price, or the second highest bid. This process was repeated at each of evaluation time points and was done after the sensory testing of the meals.

At the first evaluation time point (2 weeks of storage), no information about the two samples of jambalaya was provided before participants submitted their bids in the auction. The only information provided was the three-digit code or identification number for each meal. At the second and third evaluation time points, two pieces of information were disclosed to the participants before bidding. The order for receiving these two pieces of information was randomized among the participants. At the second time point (8 weeks of storage), the information about the name of the technology used to preserve each jambalaya sample (MAPS versus freezing) was provided to 25 participants. The information about the environmental impacts of the MAPS sample versus the frozen sample was provided to the remaining 25 participants. At the third time point (12 weeks of storage), the information disclosure was reversed. To avoid interfering with the participants' ratings of the sensory attributes of the meals, the information about the name of the technology and the environmental impacts was disclosed after the sensory testing [24].

2.2.2. Semi-Trained Panel Evaluation

Participant Selection and Orientation: A semi-trained panel ($n = 10$; 8 females, 2 males, ages 23–46) also evaluated the sensory profile of the MAPS-jambalaya and the control with rate-all-that-apply (RATA) questions. All the members of the semi-trained panel had previous experience in conducting sensory evaluation and had participated in multiple descriptive panels conducted at the WSU Sensory Science Center [22].

These evaluations were also done at Weeks 2, 8, and 12 of storage. RATA methodology has been reported to be a valid and reliable sensory profiling tool suitable for semi-trained panels [22,27]. For each session, the control trays were thawed in water at room temperature for 1.5 h. Next, each jambalaya tray (250 g) was warmed at 45–50 °C for 30 min (15 min on each side, top and bottom) with a food warmer (Glo-Ray HATCO Corporation, Milwaukee,

WI, USA). Then the trays were opened, and the jambalaya was carefully mixed. A total of 17–20 g of warmed jambalaya was then portioned into plastic cups; each sample was checked to ensure it contained all of the proportionally identical components of the jambalaya (sausage, shrimp, chicken). All samples were evaluated at 40 ± 1 °C. A 30 s break was given after the evaluation of each sample. Filtered water and unsalted crackers (Nabisco, NJ, USA) were provided as palate cleansers. Evaluations were conducted individually, in a discussion room, under white lighting.

RATA questions for the jambalaya were divided into six sections: aroma, appearance, taste/flavor, texture, mouthfeel, and aftertaste. As assessors evaluated six sensory modalities-aroma, followed by appearance, taste/flavor, texture, mouthfeel and aftertaste, they checked the terms they considered appropriate to describe the jambalaya samples (Figure 2). The list consisted of 4 to 17 terms, depending on the sensory modality. The terms used for each of the sensory modalities were defined based on pilot work. Assessors then rated the intensity of the selected terms, using a three-point structured scale (low, medium, and high). Answers were collected with a paper-based ballot. The jambalaya samples were coded with three-digit codes and presented in monadic sequential, randomized, balanced order.

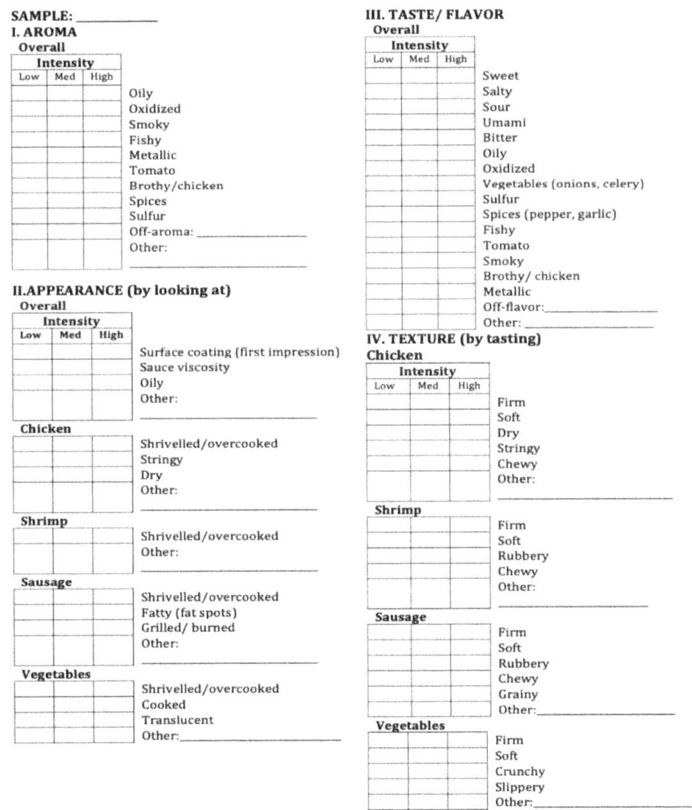

Figure 2. *Cont.*

V. MOUTHFEEL
Overall

Intensity			
Low	Med	High	
			Oily
			Greasy
			Mouthcoating
			Spicy (heat)
			Other:________________

VI. AFTERTASTE
Overall

Intensity			
Low	Med	High	
			Oily
			Bitter
			Umami
			Spices
			Heat
			Smoky
			Brothy/ chicken
			Salty
			Other:________________

Comments:________________________________

____

____

Figure 2. List of the sensory attributes tested for the MAPS-jambalaya and the control with RATA questions.

2.3. Analyses

2.3.1. Sensory Data Analysis

HUT Data: During the 12-week storage period, a repeated measures ANOVA with mixed models was conducted to evaluate the liking results of the different sensory modalities of the jambalaya samples. Processing method, storage time, having a partner and the interaction between processing method*time were analyzed as the fixed factors, with storage time as the repeated factor, and panelists as the subject factor. Means were separated with Tukey's HSD test. The JAR scale results were interpreted with penalty analysis for each of the meals at the three-evaluation time points (2, 8 and 12 weeks). XLSTAT 2017 (Addinsoft, Paris, France) statistical software was used for all sensory data analyses.

RATA Data: RATA results were analyzed by treating the RATA scores as continuous data and expanding the scale to four points (0, 1, 2, 3 for absent, low, medium, and high, respectively) [22,28,29]. A repeated measures ANOVA with mixed models was conducted. Processing method and storage time and the interaction (treatment*time) were analyzed as the fixed factors, storage time as the repeated factor, and participants as the subject factor. Means were separated with Tukey's HSD test. Significance was defined as $p < 0.05$.

If the terms were used at a frequency of 20% or less by the assessors, those terms were not considered for analysis [22]. Out of the 82 terms (Figure 2) that comprised the complete list of attributes, 14 were not considered for analysis. The results from the assessors were validated by internal agreement on the rating of the intensity of one of the attributes, as described by Montero et al. (2020) [22].

2.3.2. Online Auction Data Analysis

Data from the online auction were analyzed by a repeated measures ANOVA with mixed models. Treatment and storage time and the interaction (treatment × time) were analyzed as the fixed factors, storage time as the repeated factor, and participants as the subject factor. Means were separated with Tukey's HSD test (HSD).

3. Results and Discussion

The study evaluated two main hypotheses.

Hypothesis 1. *The acceptance/liking of different sensory characteristics of MAPS-processed jambalaya would not change significantly during storage at 2 °C as compared to a control (cooked and frozen jambalaya) over a 12-week storage period.*

Hypothesis 2. *Ecologically valid measures of consumer acceptance (a modified HUT and an online auction) would impact the degree of acceptance of the RTE meals. Hypothesis 2 employed an exploratory approach regarding how the social environment may impact the liking of RTE jambalaya meals when a partner joins the evaluation of the meals.*

3.1. HUT Evaluation

3.1.1. Comparison of Results $n = 50$ vs. $n = 71$: Inclusion of Partners

As a way to enhance ecological validity of the study and test Hypothesis 2, the responses from 21 partners were collected, included and analyzed. The obtained results were compared to the responses of the 50 participants.

In comparing the consumer liking scores of the 50 participants with those of the 71 participants (50 participants + 21 partners), there were no significant differences in the liking scores for most of the tested sensory modalities (Tables 4 and 5). These results indicate that it is reasonable to include the responses collected from the 21 partners. To have a larger number of responses in a HUT increases the robustness and power of the observed results [30]. It was observed that in certain sensory modalities such as appearance (Table 5) the p-value decreased and moved closer to being significant. This result could indicate that it was possible to identify potential difference due to the storage time effect as the number of responses increased.

Table 4. Consumer liking responses and bids' values for jambalaya MAPS-processed meals and a control (cooked and frozen meals) as evaluated by 50 and by 71 participants.

	Processing Method	Appearance	Aroma	Flavor	Texture Shrimp	Texture Chicken	Texture Sausage	Overall liking	Bids ($)
	MAPS-processed	5.66 a	5.79 a	6.00 a	5.13 a	4.95 a	5.68 a	5.61 a	3.59 a
$n = 50$	Control	5.71 a	5.92 a	5.99 b	5.35 a	4.94 a	5.70 a	5.67 a	3.48 a
	p-value	0.78	0.40	0.94	0.32	0.98	0.92	0.70	0.48
	MAPS-processed	5.63 a	5.68 a	5.88 a	5.16 a	4.93 a	5.63 a	5.34 a	-
$n = 71$	Control	5.70 a	5.85 a	5.89 a	5.27 a	4.98 a	5.77 a	5.65 a	-
	p-value	0.64	0.23	0.96	0.54	0.78	0.34	0.45	-

Different letters within a column (a, b) indicate that the tested parameter mean value was different among processing methods $p < 0.05$ as determined by using Tukey's HSD. Mean values are collapsed over participants and storage time. Results range between 1 and 7 due to the use of a 7-point hedonic scale.

Table 5. Consumer liking responses and bid values for jambalaya meals as evaluated by 50 and by 71 participants, over a 12-week storage period.

	Storage Time (Weeks)	Appearance	Aroma	Flavor	Texture Shrimp	Texture Chicken	Texture Sausage	Overall Liking	Bids ($)
	2	5.77 a	5.85 a	6.29 a	5.30 a	5.14 a	5.80 a	5.72 a	3.54 a
	8	5.69 a	5.80 a	5.80 b	5.43 a	5.00 a	5.71 a	5.59 a	3.42 a
$n = 50$	12	5.79 a	5.91 a	5.89 b	5.00 a	4.69 a	5.56 a	5.61 a	3.65 a
	p-value	0.26	0.53	<0.0001	0.03	0.02	0.20	0.46	0.50
	2	5.55 a	5.74 a	6.09 a	5.16 a	5.12 a	5.78 a	5.68 a	-
	8	5.63 a	5.72 a	5.75 b	5.41 a	5.01 a	5.73 a	5.55 a	-
$n = 71$	12	5.82 a	5.83 a	5.81 ab	5.09 a	4.75 a	5.59 a	5.54 a	-
	p-value	0.004	0.45	0.001	0.06	0.03	0.22	0.26	-

Different letters within a column (a, b) indicate the mean value of the tested parameter was different across storage times at $p < 0.05$ as determined using Tukey's HSD. Mean values are collapsed over participants and processing method. Results range between 1 and 7 due to the use of a 7-point hedonic scale.

Two main effects were evaluated with the 71 collected responses. To test Hypothesis 1, the effect of the processing method (MAPS-processed and in chilled storage vs. control, cooked and frozen storage), the effect of the storage time, and their respective interaction on the consumer liking scores for the different sensory modalities were evaluated. The interaction was not significant for any of the sensory modalities; thus, the simple effects were analyzed.

The processing method (MAPS-processed vs. control) did not significantly influence the liking scores that the participants ($n = 71$) assigned to the different sensory attributes that were evaluated (Table 4). These results indicate that the acceptance of multiple sensory attributes was comparable between the MAPS-processed meals and the cooked and frozen (control) meals.

As shown in Table 5, when the storage time effect was evaluated, only flavor liking scores decreased significantly over time ($p = 0.001$, $n = 71$). Considering the meals were evaluated on a 7-point hedonic scale, the liking score for flavor ranged between like slightly and like moderately. For a multicomponent new RTE meal that rating level can be considered as an acceptable/good liking score. On a nine-point hedonic scale, a mean liking score of 7 (like moderately) is usually indicative of highly acceptable sensory quality [31].

The results obtained from the evaluation of the processing method and the storage time indicate the potential of MAPS processing to extend the shelf life of a complex RTE meal such as the jambalaya when stored at 2 °C. Given the increased consumption of RTE nowadays, the food industry is constantly looking for alternative processing techniques that allow for the extension of the shelf of RTE meals and do not require a freezing step. Freezing has been reported as an effective method to extend the shelf life of multiple food products including RTE meals; however, it is energy intensive, and it can affect the texture-related characteristics when freeze-thawing occurs [32]. For this reason, the potential of MAPS-processing seems promising in the processing and conservation of RTE stored under refrigeration conditions.

A paucity of HUT studies have focused on the evaluation of RTE meals consumed in a home setting. A similar HUT study by Olsen et al., (2012) [13] determined that overall liking of the meal drives consumers' likelihood of buying healthy convenience meals. As in the present study, complex foods were evaluated. Two ready-to-heat meals with multiple components or different ingredients were assessed: (1) salmon fillets with raw vegetables (cauliflower, carrots, and green beans), precooked pasta and pasteurized cream and mushroom sauce; and (2) chicken breast fillets, with raw vegetables (cauliflower, carrots, and green beans), precooked white rice and pasteurized red bell pepper sauce. However, Olsen's results focused on drivers of overall liking including appearance, odor, amount of ingredients, and flavor. While these attributes were also considerations in our study, our study also focused on the effects of the processing method, storage time and the enhanced social environment in which the food was consumed along with the willingness to purchase of the product.

3.1.2. Willingness to Pay as Evaluated with an Online Auction

To evaluate the participants willingness to pay for the jambalaya meals an online auction was conducted. The auction protocol established that participants received money for placing the bids at each evaluation time point, and because the economic resources were limited, only the 50 participants joined the online auction. The bidding was performed based on the sensory evaluation of each of the jambalaya meals. Information of the processing system was not included at the time of placing the bid at the first time point; this type of information was given at the second and third time points. However, during these two points, the information was provided after the participants conducted the sensory evaluation and prior to submitting their bids [24]. As previously mentioned, the effect of the processing method, storage time and their interaction on the bid values were also analyzed. The interaction was not significant and for this reason the main effects were interpreted. The bid values did not significantly differ ($p = 0.48$) between the MAPS-processed jambalaya and the control, and they were not influenced by the storage time (Table 4).

The bids values assigned by the participants to the meals were comparable to commercially available jambalaya meals. The mean bid values ranged from \$3.48–3.74 for the MAPS-processed jambalaya and from \$3.33–3.56 for the control.

3.1.3. Effect of Eating with Having a Partner on the Consumer Liking Scores of the Jambalaya Meals

To test Hypothesis 2, the effect of eating with a partner in the liking scores of the sensory characteristics of the meals were also evaluated (Table 6).

Table 6. Consumer liking responses and bid values for jambalaya meals as evaluated by participants who had or did not have a partner ($n = 50$).

Partner	Appearance	Aroma	Flavor	Texture Shrimp	Texture Chicken	Texture Sausage	Overall Liking	Bids (\$)
Yes	5.88 a [1]	5.98 a	6.04 a	5.40 a	5.07 a	5.83 a	5.69 a	3.59 a
No	5.54 b	5.76 a	5.96 a	5.13 a	4.85 a	5.59 a	5.60 a	3.50 a
p-value	0.04	0.18	0.63	0.24	0.35	0.20	0.62	0.57

[1] Different letters within a column (a,b) indicate that the tested parameter mean value was different among those participants with a partner vs. those without at $p < 0.05$ as determined using Tukey's HSD. Data are collapsed over processing method and storage time. Results range between 1 and 7 due to the use of a 7-point hedonic scale.

It was determined that those participants having a partner gave a significantly higher score ($p = 0.04$) to the appearance of the jambalaya meals. The value was 5.88 vs. 5.54 for those participants without a partner. This liking value is associated with a rating between like slightly and like moderately on a seven-point hedonic scale. Overall, there was a trend in the liking scores of the evaluated sensory attributes; those participants with a partner gave higher scores to the liking of all the evaluated sensory attributes of the meals. Laureati and Pagliarini (2019) [5] defined three main contextual factors that influence eating behavior when conducting consumer testing, the meal (i.e., sensory characteristics); the physical environment (i.e., appropriate location and setting); and the social environment or social interaction (i.e., people present at the experiment). In this study, each of those factors was explored and the social environment seemed to be positively impacted by the partners addition. Piliner, Bell, Kinchla and Hirsch (2003) [33] stated that social interaction has a positive effect on food consumption of naturally created groups but not artificially created. In our study, the partners (spouse, friend, roommate) could be categorized as members of a naturally created group for the participants or that will evoke a more realistic consumption situation [34,35].

Petit and Siefferman (2007) [36] maintain that conducting food testing in naturalistic conditions is more advantageous than in-lab tests due to the realism of the evaluation, but situational tests such as HUTs can be more expensive and time-consuming than in-lab ones. As shown in our study, the addition of a partner could mitigate some of these downsides, mainly the one related to costs. Currently, with COVID-19 restrictions, including responses from partners could represent a simple option for sensory scientists and food companies to increase the number of respondents and enhance ecological validity of the study.

In this study the modified HUT and an online auction seemed to positively impact the degree of acceptance of the RTE meals.

To address Hypothesis 2, on the last evaluation time point the 50 participants were asked to provide feedback about their experience participating in the HUT. The comments were carefully reviewed and divided into seven categories (Table 7). Almost 30% of the participants of the study indicated that they had a positive experience when sharing the evaluation of the meals with their partners. One of the participants mentioned "sharing the samples with my partner is enjoyable, because after we record our ratings individually, we compare and discuss the two samples".

Table 7. Categories, frequency of mention and some comment examples from the participants of the study (*n* = 50).

Categories	Frequency of Mention (%)	Comment Examples
Enjoyed experience with partner	26	"I enjoyed sharing this with my partner, who was equally eager to partake due to being unable to partake in the in-lab studies" "Nice to do the tasting a home on my own time and get feedback from my wife" "Sharing the samples with my partner is enjoyable, because after we record our ratings individually, we compare and discuss the two samples" "I enjoyed having this experience with my partner as we could discuss what we liked and didn't like"
HUT vs. in-lab evaluation	42	"Way more convenient than in-lab and it's the same setting in which I'd be eating the convenience meal if I purchased one" "I liked being able to utilize my samples as dinner on each of the nights. I felt that I could give a better evaluation because I had more of a sample to taste" "I really appreciate the research these students are involved in and the fact they incorporate real people and samples rather than all lab studies" "I like bringing the samples home to taste because it is more realistic and relevant to how I would actually eat the food"
Time flexibility	14	"This in-home study offered me more flexibility with the time period. I didn't feel pressured to get out of my chair and let the next person in line have a seat" "I spend more time evaluating at home versus in the lab where it is hurried to leave" "I liked being able to take my time and enjoy each sample without feeling rushed" "I enjoyed having more flexibility in the amount of time I could take the survey"
Fun/positive experience	42	"My husband and I had fun with your in-home study, thank you" "It was a pleasant in-home study, and the quality was good" "I enjoyed being able to do this at home" "This was fun to participate in and it was easy"
Liking of the meals	16	"I very much enjoyed the samples and will be sad to not have them anymore" "I enjoyed the food and getting cash for participating" " . . . the jambalaya was something I looked forward to, I didn't get tired of it" "I could have eaten both samples myself"
Willingness to pay for the meals	8	"We both liked the meals and would definitely purchase if available at the store" "I hope I will see the samples in the store" "I could better consider if the sample would be something I would purchase" "We would be willing to spend a little more on these since they have a large variety in them . . . "

The category HUT vs. in-lab evaluation shows how over 40% of the participants preferred doing the sensory evaluation of the jambalaya meals at home vs. at the SST or in-lab set up. One of the participants mentioned "I like bringing the samples home to taste because it is more realistic and relevant to how I would actually eat the food"; this comment points to the value of conducting the sensory evaluation under a setup that is more realistic and familiar to the participant. These aspects contribute to the enhancement of the ecological validity of the study [1,36]. Time flexibility was another topic mentioned by the participants in their comments. Almost 15% of their comments indicated that the participants liked being able to take their time and enjoy each sample without feeling rushed. This is another positive component of the in-home evaluation setup. Another category of frequently mentioned comments (42%) was that the participants had a fun or positive experience while participating in the study.

The two final categories were liking of the meal and willingness to pay-related comments. Almost 20% of the participants mentioned liking the meals and almost 10% expressed their willingness to pay for the jambalaya meals if they were available in the market.

Based on the type of comments mentioned by the participants of the HUT, it seems that overall, the HUT was a pleasant, positive experience that allowed them to manage the time for evaluating the meals at their own convenience. This seems like a promising way to accomplish Hypothesis 2.

3.1.4. Potential Improvements on the Jambalaya RTEs Meals

To evaluate the impact of the processing methods and storage time on the spiciness and texture perception of the meat components of the meals, JAR questions were used. The main results are presented in Figure 3.

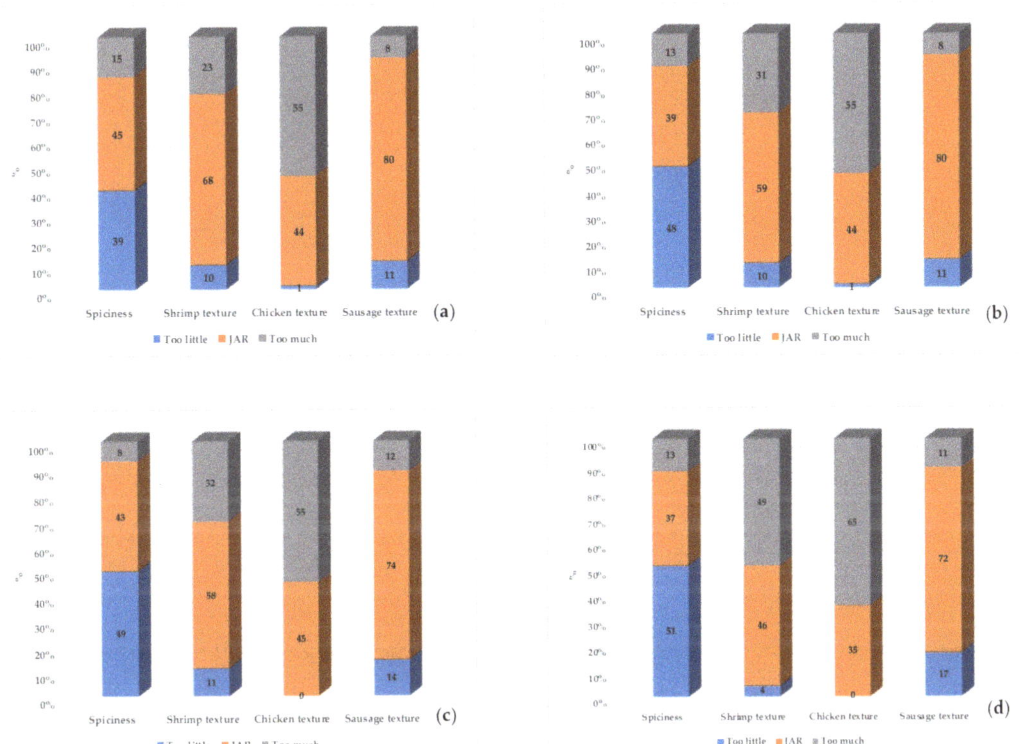

Figure 3. Percentage of responses for the indicated anchor of the JAR scale in the sensory evaluation of the intensity of spiciness and texture of the three meat components of the jambalaya meals (MAPS-processed and control) over a 12-week storage period (*n* = 71). (**a**) Control after 2 weeks of storage; (**b**) Control after 12 weeks of storage; (**c**) MAPS-jambalaya after 2 weeks of storage; (**d**) MAPS-jambalaya after 12 weeks of storage.

JAR questions are useful as they can provide focused direction to new product development. With penalty analysis, it is possible to determine which elements most impact the overall liking of a product.

The spiciness intensity in most of the samples was considered not spicy. The texture of the shrimp and sausage were mostly rated "just-about-right" on the JAR scale for both the control and the MAPS-processed jambalaya. In addition, when the first evaluation time point (2 weeks) was compared to the last one (12 weeks), the ratings were similar. Chicken texture was the one attribute most penalized by the participants of the study as they considered it to be too chewy/overcooked. This was observed for both the control and the MAPS-processed jambalaya in both time points, after 2 and 12 weeks of storage, respectively.

To complement the potential improvements that could be made to the jambalaya meals, on the last evaluation time point the participants (n = 50) answered a series of JAR questions.

Most of the aspects of the meals were rated close to two on the three-point scale, which corresponds to the JAR point. The other two values on the scale were 1 (=less than I would like) and 3 (=more than I would like).

The portion size was scored with an average value of 1.68, the quantity of sauce was rated with 1.96, the quantity and size of vegetables were scored with 1.56 and 1.82, respectively. The quantity of shrimp, chicken and sausage with 1.74, 2.12 and 1.72. The saltiness was rated with a value of 2.04, also very close to the JAR point.

The only question that pointed out a potential improvement was the question related to removing the tails of the shrimps (score = 2.84). Most of the participants favored the removal of the tails.

Overall, JAR questions contributed on having a better understanding of Hypothesis 1.

3.2. Semi-Trained Panel Evaluation

The other evaluation was done by a semi-trained panel. This panel used RATA questions as the sensory evaluation technique. This component of the study also supported Hypothesis 1 of the study.

For the analysis, the effect of the treatment (MAPS-processed vs. control), the effect of the storage time, and their respective interaction on the intensity of different sensory attributes were evaluated. The interaction was not significant for any the sensory modalities; thus, the simple effects were analyzed.

As shown in Table 8, only the storage significantly impacted the intensity of different sensory attribute scores as evaluated by the panelists.

Table 8. Mean intensity values of multiple sensory attributes of jambalaya meals as evaluated by a semi-trained panel (n = 10) over a 12-week storage period with RATA.

Storage Time (Weeks)	Oxidized Aroma	Brothy-Chicken Aroma	Oily Appearance	Shriveled/Overcooked Chicken Appearance	Chewy Sausage
2	0.05 a	2.15 a	0.25 a	2.45 a	1.55 a
8	0.10 b	2.85 b	1.50 b	1.50 b	2.40 b
12	0.68 ba	2.60 ab	1.72 b	1.15 b	2.01 ab
p-value	<0.0001	0.002	<0.0001	<0.0001	0.001

Different letters within a column (a,b) indicate that attribute intensities were different among storage times at $p < 0.05$ as determined using Tukey's HSD. These results range between 0 and 3 due to the use of a four-point scale of 0, 1, 2, 3 (absent, low, medium, high). Mean values are collapsed over processing method, replicate and panelists.

The aroma related attributes, for example oxidized and brothy-chicken significantly increased as storage time increased. In the case of oxidized aroma, the values ranges were less than 1; therefore, they are considered to be more in the low range. The brothy-chicken aroma increased mainly after 8 weeks of storage. Barnett et al. (2019) [23] reported increases in the intensities of some aroma and flavor attributes of microwave-processed Cajun chicken pasta meals as the storage time increased. The concentration of aroma and flavor may have resulted from potential water migration from the package [23,37].

For the appearance related attributes, oily appearance intensity significantly increased due to the effect of the storage time, meanwhile the shriveled/overcooked appearance in chicken decreased. This result could be explained by the fact that this specific attribute could be a difficult one to evaluate, therefore the panelists gave initially medium to high score and then after 8 weeks of storage a score between low and medium. The opposite trend could have been expected, that as an effect of the potential water migration in the tray [23,37] the chicken experienced some level of dehydration and therefore was perceived by the panelists as more shriveled-overcooked.

Another appearance attribute that was significantly different was the shriveled-overcooked appearance of the shrimp. In this case the interaction processing method × storage time was significant. At 2 weeks of storage the MAPS-processed jambalaya (2.20) was rated significantly higher ($p = 0001$) as the control (0.80). This result could be explained by the fact that the MAPS-processed meal goes through an extra thermal processing step. However, at 8 and 12 weeks of storage there were not significant differences when comparing the MAPS-processed jambalaya to the control. The intensities of shriveled-overcooked appearance of the shrimp at 8 and 12 weeks of storage ranged between 2.00 and 1.50.

It is also important to consider that these two texture-related appearance attributes, shriveled/overcooked appearance in chicken and shriveled-overcooked appearance of the shrimp could have more impact if they significant difference in the texture sensory modality and that was not the case.

The final type of sensory attribute that was impacted by storage time was the texture of the sausage, specifically its chewiness. This attribute significantly increased as storage time increased primarily when Week 2 and Week 8 of storage were compared.

Overall, based on the RATA results few sensory attributes were affected by the main factors evaluated in the study.

4. Conclusions

Our findings showed that the acceptance/liking of different sensory characteristics of MAPS-processed jambalaya did not change significantly during storage at 2 °C as compared to a control (cooked and frozen jambalaya stored at −31 °C) over a 12-week storage period. Using a HUT to evaluate consumer acceptance of jambalaya and including partner participation, is a promising way of testing acceptance in a more realistic context. An on-line auction with a HUT sensory testing of jambalaya meals showed that consumers are willing to pay a price that is comparable to commercially available jambalaya meals. Both were ecologically valid measures of consumer acceptance and positively impacted the degree of acceptance of the RTE meals.

This study has utility and presents an innovative approach as compared to previously reported HUTs. The inclusion of a partner in the evaluation of a ready-to-eat meal processed with a novel technology, such as microwave-assisted pasteurization has not been previously reported. In addition, the evaluation was conducted during an extended period of time of 12 weeks. Most reported HUTs are conducted within a week and the exposure to the product being tested occurs one time or consecutive times within a short period of time (e.g., one week). In addition, the study incorporated the evaluation of a complimentary evaluation with a semi-trained panel that used a rapid method, rate-all-that-apply. These elements were combined with an online auction to determine consumers' willingness to pay for the meals.

As for the HUT results, it should be noted that this study had an observational/exploratory approach, focused on a specific RTE meal. Because of the relatively restricted number of consumers that joined the study, future research should evaluate the effect of the inclusion of partner as an intentional/designated treatment and target a larger number of participants. A HUT and an SST (laboratory test or central location test) could be conducted simultaneously at a defined storage period and the results of both tests compared.

In addition, the inclusion of an analytical technique, like gas chromatography-mass spectrometry and texture profile analysis (TPA), could be considered for future studies to describe specific chemical and physical changes in the meals. However, multicomponent meals are a complex/challenging food matrix to work with that is why sensory evaluation seems like a more effective methodology to characterize them during storage instead of analytical techniques.

Author Contributions: Conceptualization, M.L.M., D.G., C.F.R. and R.K.G.; methodology, M.L.M., D.G., C.F.R. and R.K.G.; formal analysis, M.L.M.; investigation, M.L.M. and D.G.; data curation, M.L.M.; writing—original draft preparation, M.L.M. and C.F.R.; writing—review and editing, M.L.M., C.F.R., D.G., R.K.G. and J.T.; supervision, C.F.R. and R.K.G.; project administration, C.F.R. and R.K.G.; funding acquisition, J.T. All authors have read and agreed to the published version of the manuscript.

Funding: This research was funded by the USDA AFRI Grant 2016-68003-24840. This work was partially supported by the USDA National Institute of Food and Agriculture, Hatch project with Accession #1016366.

Institutional Review Board Statement: The WSU Institutional Review Board approved this study for conducting tests with human subjects, under the title Consumer Preferences of Jambalaya IRB #16994.

Informed Consent Statement: Informed consent was obtained from all subjects involved in the study.

Data Availability Statement: The data that support the findings of this study are available from the corresponding author upon reasonable request. The data are not publicly available due to privacy issues.

Acknowledgments: The authors would like to acknowledge the panelists that joined the study, the staff and the students from Washington State University that collaborated in the preparation and processing of the jambalaya meals. Special thanks to Elizabeth Siler of the Department of English at Washington State University for her contribution in editing and reviewing the manuscript. Maria Laura Montero acknowledges doctoral fellowship support from the University of Costa Rica.

Conflicts of Interest: The authors declare no conflict of interest. The funders had no role in the design of the study; in the collection, analyses, or interpretation of data; in the writing of the manuscript, or in the decision to publish the results.

References

1. Stelick, A.; Dando, R. Thinking outside the booth—The eating environment, context and ecological validity in sensory and consumer research. *Curr. Opin. Food. Sci.* **2018**, *21*, 26–31. [CrossRef]
2. Koster, E.G. The psychology of food choice: Some often encountered fallacies. *Food Qual. Prefer.* **2003**, *14*, 359–373. [CrossRef]
3. Hartson, R.; Pyla, P.S. *UX Book: Process and Guidelines for Ensuring a Quality User Experience*; Morgan Kaufmann Publishers Inc.: San Francisco, CA, USA, 2012; pp. 503–536. [CrossRef]
4. Thomas, J.C.; Kellogg, W.A. Minimizing ecological gaps in interface design. *IEEE Softw.* **1989**, *6*, 78–86. [CrossRef]
5. Laureati, M.; Pagliarini, E. The effect of context on children's eating behavior. In *Context: The Effects of Environment on Product Design and Evaluation*; Meiselman, H.L., Ed.; Woodhead Publishing: Cambridge, UK, 2019; pp. 287–305. [CrossRef]
6. Zhang, M.T.; Jo, Y.; Lopetcharat, K.; Drake, M.A. Comparison of a central location test versus a home usage test for consumer perception of ready-to-mix protein beverages. *J. Dairy Sci.* **2020**, *103*, 3107–3124. [CrossRef] [PubMed]
7. Boutrolle, I.; Delarue, J.; Arranz, D.; Rogeaux, M.; Köster, E.P. Central location test vs. home use test: Contrasting results depending on product type. *Food Qual. Prefer.* **2007**, *18*, 490–499. [CrossRef]
8. Lawless, H.T.; Heymann, H. *Sensory Evaluation of Food: Principles and Practices*, 2nd ed.; Springer: New York, NY, USA, 2010; pp. 351–355. [CrossRef]
9. Hein, K.A.; Hamid, N.; Jaeger, S.R.; Delahunty, C.M. Effects of evoked consumption contexts on hedonic ratings: A case study with two fruit beverages. *Food Qual. Prefer.* **2012**, *26*, 35–44. [CrossRef]
10. Clendenen, V.I.; Herman, C.P.; Polivy, J. Social facilitation of eating among friends and strangers. *Appetite* **1994**, *23*, 1–13. [CrossRef] [PubMed]
11. de Graaf, C.; Cardello, A.V.; Mathew Kramer, F.; Lesher, L.L.; Meiselman, H.L.; Schutz, H.G. A comparison between liking ratings obtained under laboratory and field conditions: The role of choice. *Appetite* **2005**, *44*, 15–22. [CrossRef]
12. Sveinsdóttir, K.; Martinsdóttir, E.; Thórsdóttir, F.; Schelvis, R.; Kole, A.; Thórsdóttir, I. Evaluation of farmed cod products by a trained sensory panel and consumers in different test settings. *J. Sens. Stud.* **2010**, *25*, 280–293. [CrossRef]
13. Olsen, N.V.; Menichelli, E.; Sørheim, O.; Næs, T. Likelihood of buying healthy convenience food: An at-home testing procedure for ready-to-heat meals. *Food Qual. Prefer.* **2012**, *24*, 171–178. [CrossRef]
14. French, S.A. Pricing effects on food choices. *J. Nutr.* **2003**, *133*, 841S–843S. [CrossRef]
15. Ramirez, M.; Lopez, A.; Hernandez, C.; Lavios, J.J.; Poza, D.; Ranieri, A.; Herranz, R. Review of auction-based markets. In *Application of Agent-Based Computational Economics to Strategic Slot Allocation (ACCESS)*; 2014; pp. 1–61. Available online: https://www.researchgate.net/publication/322276880_Review_of_Auction-Based_Markets (accessed on 15 March 2021).
16. Breidert, C.; Hahsler, M.; Reutterer, T. A review of methods for measuring willingness-to-pay. *Innov. Mark.* **2006**, *2*, 8–32.
17. Huang, L.; Hwang, C.-A. In-package pasteurization of ready-to-eat meat and poultry products. In *Advances in Meat, Poultry and Seafood Packaging*; Kerry, J.P., Ed.; Woodhead Publishing: Cambridge, UK, 2012; pp. 437–450. [CrossRef]
18. Okrent, A.M.; Kumcu, A.U.S. Households' Demand for Convenience Foods. United States Department of Agriculture, Economic Research Service, ERR-211. 2016. Available online: https://www.ers.usda.gov/webdocs/publications/80654/err-211.pdf?v=0 (accessed on 15 March 2021).
19. Tang, J.; Hong, Y.K.; Inanoglu, S.; Liu, F. Microwave pasteurization for ready-to-eat meals. *Curr. Opin. Food. Sci.* **2018**, *23*, 133–141. [CrossRef]

20. Auksornsri, T.; Songsermpong, S. Lethality and quality evaluation of in-packaged ready-to-eat cooked Jasmine rice subjected to industrial continuous microwave pasteurization. *Int. J. Food Prop.* **2017**, *20*, 1856–1865. [CrossRef]
21. Tang, J. Unlocking Potentials of Microwaves for Food Safety and Quality. *J. Food Sci.* **2015**, *80*, E1776–E1793. [CrossRef]
22. Montero, M.L.; Sablani, S.; Tang, J.; Ross, C.F. Characterization of the sensory, chemical, and microbial quality of microwave-assisted, thermally pasteurized fried rice during storage. *J. Food Sci.* **2020**, *85*, 2711–2719. [CrossRef] [PubMed]
23. Barnett, S.M.; Sablani, S.S.; Tang, J.; Ross, C.F. Utilizing herbs and microwave-assisted thermal sterilization to enhance saltiness perception in a chicken pasta meal. *J. Food Sci.* **2019**, *84*, 2313–2324. [CrossRef] [PubMed]
24. Garrido, D.; Gallardo, R.K.; Ross, C.F.; Montero, M.L.; Tang, J. The effect of intrinsic and extrinsic quality on the willingness to pay for a convenient meal: A combination of home-use-test with online auctions. *J. Sens. Stud.* **2021**, e12682. [CrossRef]
25. Sloan, E.A. The top 10 food trends of 2021. *Food Technol.* **2021**, *75*, 23–35.
26. Lusk, J.L.; Shogren, J.F. *Experimental Auctions: Methods and Applications in Economic and Marketing Research*; Cambridge University Press: Cambridge, UK, 2007; pp. 62–94. [CrossRef]
27. Giacalone, D.; Hedelund, P.I. Rate-all-that-apply (RATA) with semi-trained assessors: An investigation of the method reproducibility at assessor, attribute and panel level. *Food Qual. Prefer.* **2016**, *51*, 65–71. [CrossRef]
28. Meyners, M.; Jaeger, S.R.; Ares, G. On the analysis of Rate-All-That-Apply (RATA) data. *Food Qual. Prefer.* **2016**, *49*, 1–10. [CrossRef]
29. Vidal, L.; Ares, G.; Hedderley, D.I.; Jaeger, S.R. Comparison of rate-all-that-apply (RATA) and check-all-that-apply (CATA) questions across seven consumer studies. *Food Qual. Prefer.* **2018**, *67*, 49–58. [CrossRef]
30. Hough, G.; Wakeling, I.; Mucci, A.; Chambers, E.; Méndez Gallardo, I.; Rangel Alves, L. Number of consumers necessary for sensory acceptability tests. *Food Qual. Prefer.* **2006**, *17*, 522–526. [CrossRef]
31. Everitt, M. Consumer-Targeted Sensory Quality. In *Global Issues in Food Science and Technology*; Barbosa-Cánovas, G., Mortimer, A., Lineback, D., Spiess, W., Buckle, K., Colonna, P., Eds.; Academic Press: Cambridge, MA, USA, 2009; pp. 117–128. [CrossRef]
32. Yu, L.; Turner, M.S.; Fitzgerald, M.; Stokes, J.R.; Witt, T. Review of the effects of different processing technologies on cooked and convenience rice quality. *Trends. Food Sci. Technol.* **2017**, *59*, 124–138. [CrossRef]
33. Pliner, P.; Bell, R.; Kinchla, M.; Hirsch, E. Time to eat? The impact of time facilitation and social facilitation on food intake. In Proceedings of the Pangborn Sensory Science Symposium Abstract Book, Boston, MA, USA, 20–24 July 2003.
34. Di Monaco, R.; Giacalone, D.; Pepe, O.; Masi, P.; Cavella, S. Effect of social interaction and meal accompaniments on acceptability of sourdough prepared croissants: An exploratory study. *Food Res. Int.* **2014**, *66*, 325–331. [CrossRef]
35. King, S.C.; Weber, A.J.; Meiselman, H.L.; Lv, N. The effect of meal situation, social interaction, physical environment and choice on food acceptability. *Food Qual. Prefer.* **2004**, *15*, 645–653. [CrossRef]
36. Petit, C.; Siefferman, J.M. Testing consumer preferences for iced-coffee: Does the drinking environment have any influence? *Food Qual. Prefer.* **2007**, *18*, 161–172. [CrossRef]
37. Zhang, H.; Tang, Z.; Rasco, B.; Tang, J.; Sablani, S.S. Shelf-life modeling of microwave-assisted thermal sterilized mashed potato in polymeric pouches of different gas barrier properties. *J. Food Eng.* **2016**, *183*, 65–73. [CrossRef]

Article

Exploring the Effects of Immersive Virtual Reality Environments on Sensory Perception of Beef Steaks and Chocolate

Emily Crofton [1,*], Niall Murray [2] and Cristina Botinestean [1]

[1] Teagasc Food Research Centre, Department of Food Quality and Sensory Science, Dublin D15 KN3K, Ireland; cristina.botinestean@teagasc.ie
[2] Athlone Institute of Technology, Faculty of Engineering and Informatics, Department of Computer and Software Engineering, Athlone N37 F6D7, Ireland; nmurray@research.ait.ie
* Correspondence: Emily.Crofton@teagasc.ie

Abstract: Virtual reality (VR) technology is emerging as a tool for simulating different eating environments to better understand consumer sensory response to food. This research explored the impact of different environmental contexts on participants' hedonic ratings of two different food products: beef steaks, and milk chocolate, using VR as the context-enhancing technology. Two separate studies were conducted. For beef, two different contextual conditions were compared: traditional sensory booths and a VR restaurant. For chocolate, data were generated under three different contextual conditions: traditional sensory booths, VR Irish countryside; VR busy city (Dublin, Ireland). All VR experiences were 360-degree video based. Consumer level of engagement in the different contextual settings was also investigated. The results showed that VR had a significant effect on participants' hedonic responses to the food products. Beef was rated significantly higher in terms of liking for all sensory attributes when consumed in the VR restaurant. While for chocolate, the VR countryside context generated significantly higher hedonic scores for flavour and overall liking in comparison to the sensory booth. Taken together, both studies demonstrate how specific contextual settings can impact participants' sensory response to food products, when compared to a traditional sensory laboratory condition.

Keywords: beef; chocolate; acceptability; virtual reality; immersive environments

Citation: Crofton, E.; Murray, N.; Botinestean, C. Exploring the Effects of Immersive Virtual Reality Environments on Sensory Perception of Beef Steaks and Chocolate. *Foods* **2021**, *10*, 1154. https://doi.org/10.3390/foods10061154

Academic Editor: Damir Torrico

Received: 1 April 2021
Accepted: 12 May 2021
Published: 21 May 2021

Publisher's Note: MDPI stays neutral with regard to jurisdictional claims in published maps and institutional affiliations.

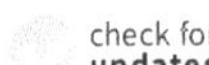

1. Introduction

The act of eating is among the most pleasurable and enjoyable experiences in our everyday lives. It is also a highly multisensory experience involving a dynamic interplay between the brain and each of the five senses, comprising visual and auditory cues, smells, tastes, and tactile sensations. While much of the pleasure we perceive from food depends on the quality of the ingredients and how they are prepared, our surrounding environment or context in which we consume the food also plays a critical role. The location or setting, the type of music, lighting or sounds, and the presence or not of family and friends, are all examples of contextual factors that consumers typically consider when making sensory judgements about a food or a meal [1–3].

Despite consumption context playing an important role in how people perceive food, consumer-based sensory tests typically take place in a highly controlled sensory laboratory setting, which lacks environmental or contextual cues. While a sensory testing facility is specifically designed to minimise potential variation caused by the external environment, the setting does not represent how consumers interact with food products in real life [4]. In addition, and beyond contextual cues, a sterile sensory laboratory setting may also induce feelings of boredom, disinterest, and a lack of engagement among participants, resulting in data that poorly predicts the success rate of newly launched products in the marketplace.

Although many factors can contribute to the poor predictive validity of consumer sensory data, researchers have cited an overall lack of effective engagement by study participants during the sensory task as a contributing cause [2,5,6]. As a result, there is considerable interest among researchers in investigating ways to evoke consumption contexts during the sensory evaluation of foods, with the aim of improving the ecological validity of consumer data. To date, several different approaches have been proposed including written, imagined or image depicted scenarios [4,7], the creation of an immersive environment, such as a beach [8] or café [9], and the use of virtual reality technology [10–12].

The use of virtual reality (VR) technology to simulate immersive contextual settings for hedonic testing is becoming an increasingly popular study area in sensory science (see Crofton et al. [13] for a review describing recent advancements in virtual and augmented reality and their potential application in sensory science). Research comparing consumer responses generated in immersive VR environments and sensory laboratory settings have been recently reported for a range of food products, including chocolate [11], tea break snacks [14], cheese [15], and alcoholic beverages [10,12]. Consumer hedonic ratings were found to change depending on the context for some products [12,15], but not others [10,11], although this may be explained by the different environmental contexts used across these studies. Outside of academia, the food industry are exploring ways VR technology can be used to customise a unique multisensory experience for their consumers. For instance, Guinness, a popular Irish stout sold worldwide, recently designed specific VR environments to complement the sensory profile of a new range of beers [16], while restaurants are adopting the technology to create more memorable and engaging food experiences [17].

The potential for VR technology to generate ecologically valid data comparable to that obtained in a natural consumption setting, in a faster, more cost-effective and controlled experimental manner, is a promising innovation strategy for the food industry to consider [14,18]. In this regard, the overall aim of this research was to explore the effect of different environmental contexts on consumers' hedonic responses to two different food products: beef, and milk chocolate, using VR as the context-enhancing technology. Beef and chocolate are associated with very different types of eating occasions. For instance, beef is typically eaten in the context of a meal, while chocolate is considered a snack suited to a range of eating occasions [11,19]. As certain products are better suited to specific situations of consumption [20], our hypothesis was that a virtual environment appropriate to each product, may enhance consumer hedonic ratings, when compared to sensory testing booths as the control setting. Therefore, two separate studies were conducted. For beef, two different contextual conditions were compared; (1) traditional sensory booth; (2) VR restaurant. For chocolate, data were generated under three different contextual conditions; (1) traditional sensory booth; (2) VR Irish countryside; (3) VR busy city (Dublin, Ireland). Consumer level of engagement in the different contextual settings was also investigated.

2. Materials and Methods

2.1. Participants

Participants for both the beef trial (n = 30, 15 females and 15 males, ranging in age from 22 to 55 years) and the chocolate trial (n = 30, 18 females and 12 males, ranging in age from 22 to 65 years) were recruited via internal email from a pool of staff and students based at Teagasc Food Research Centre, Dublin, Ireland. Participants had to meet the following inclusion criteria: (1) consumed the product used in the study at least twice per month; (2) had no known history of food allergies related to the study products; and (3) did not suffer from motion sickness. Selected participants were asked to refrain from eating, drinking, or smoking for at least 1 h prior to the start of the trial. Informed consent was obtained from each participant and they were free to withdraw from the study at any time. The sensory trials were conducted in accordance with the guidelines for ethical and professional practices for the sensory analysis of foods as set out by the Institute of Food Science and Technology [21] and the American Meat Science Association [22]. The sessions were conducted during mid-morning (10:00–11:30 a.m.) and took place at the Sensory

Science Suite at Teagasc Food Research Centre, which has been designed in accordance with ISO 8586:2007.

2.2. Product Stimuli and Sensory Procedure

2.2.1. Beef Trial

Individual beef steaks of 2.54 cm thickness were cut from the M. longissimus lumborum. Each beef steak was vacuum-packaged and stored at $-20\ ^{\circ}\text{C}$ until analysis. The steaks were placed in a refrigerator to thaw 24 h prior to sensory analysis. One hour prior to cooking, the steaks were taken from the refrigerator and removed from the vacuum bag. Steaks were cooked on a Velox grill using minimal cooking oil and no seasoning, until an internal temperature of 71 $^{\circ}\text{C}$ was reached [22]. Each steak was removed from the grill, allowed to rest for 2 min, wrapped in aluminium foil and labelled with a random three-digit code. Each steak was cut into two equal-sized portions and presented monadically to participants in two different environments: a traditional sensory booth and a virtual reality restaurant. The presentation order of the environmental context was completely balanced across participants (i.e., 15 participants assessed the beef in the traditional booth first, and 15 in the virtual reality restaurant first). Participants were instructed to evaluate the meat for liking in terms of smell, tenderness, juiciness, beef flavour, and overall liking on a 9-point hedonic scale, where 1 = dislike extremely and 9 = like extremely. In an effort to retain the level of immersion in the virtual reality condition, participants provided answers to the questions verbally in all conditions, and the data was immediately recorded by the researchers via Compusense Cloud® (West Guelph, ON, Canada). All participants were familiarised with the questionnaire design and testing procedure prior to starting the trial and were instructed on how to wear and use the VR headsets. Beef samples were served to participants after they were wearing the HMD device, and researchers were available to assist the participants, if required. A five-minute break was enforced between each tasting condition and plain crackers (Carr's, UK) and mineral water (Evian, France) were provided as palate cleansers to avoid product carry-over effects.

2.2.2. Chocolate Trial

Milk chocolate (20% cocoa, Cadbury's) was the sensory stimuli used in this trial. It was purchased at a local convenience store and was stored at room temperature prior to the sensory trial. Three identical square pieces of chocolate, taken from the same chocolate bar, were placed on separate ceramic white plates 5 min prior to each testing session, and presented monadically to participants in three testing environments—a traditional sensory booth, a VR Irish countryside, and a VR busy city. The presentation order of the testing environment was completely balanced across participants. Participants were instructed to evaluate the chocolate for liking in terms of smell, flavour, sweetness, texture, smoothness, and overall liking on a 9-point hedonic scale, where 1 = dislike extremely and 9 = like extremely. Similar to the approach taken in the beef trial, participants provided answers to the questions verbally in all conditions in an effort to retain the degree of perceived immersion in the VR session. All participants were familiarised with testing procedures prior to starting the test. Data was recorded by the researchers via Compusense Cloud ® (West Guelph, ON, Canada). Plain crackers (Carr's, UK) and mineral water (Evian, France) were used as palate cleansers and a five-minute break was enforced between each tasting condition to avoid product carry-over effects.

2.3. Contextual Settings

2.3.1. Beef Trial

Two different experimental conditions were set-up; a traditional sensory booth (Figure 1a) and an immersive VR restaurant (Figure 1b). The VR restaurant context was captured using a Samsung Gear 360 4 K Ultra High Definition (HD) camera at a restaurant in Dublin, Ireland. The restaurant was open to customers during the recording to ensure the atmosphere created was as close as possible to real-life conditions. Audio recordings consisting of indistinguishable conversation and background noises were also recorded. The 360-degree video was 2 min in length and was set to loop throughout the tasting condition. The video was presented to participants through a head mounted display (HMD) (Oculus Go).

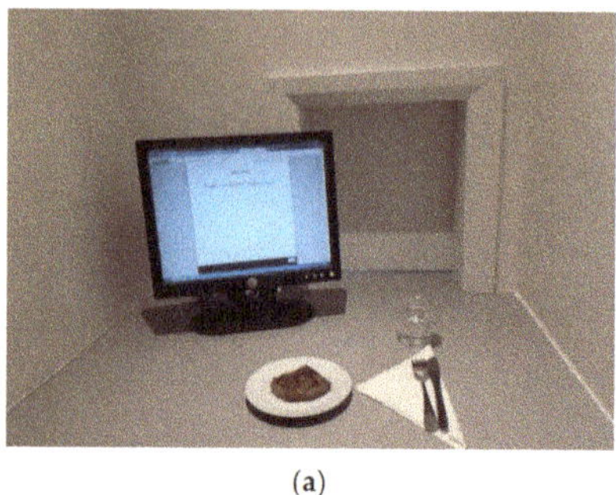

(a)

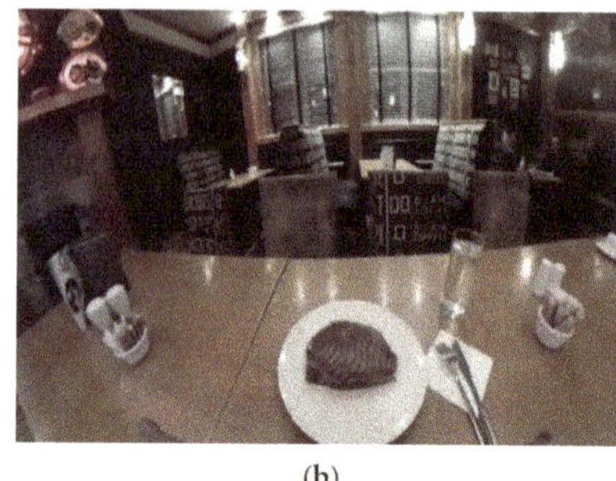

(b)

Figure 1. Contextual settings for the sensory evaluation of beef samples: (**a**) Traditional sensory booth; (**b**) Virtual Reality (VR) restaurant.

2.3.2. Chocolate Trial

Participants evaluated chocolate in three different environmental contexts: a traditional sensory booth, a VR busy city (Figure 2a), and a VR Irish countryside (Figure 2b). These specific VR contexts were chosen as they were expected to elicit different hedonic responses from participants. The VR environments were created using 360-degree videos displayed through a HMD (Oculus Go). The VR videos were captured using a Garmin VIRB 360 camera and included audio recordings. Both 360-degree videos had a length of 2 min and the videos were set to loop during each tasting session. The 360-degree busy city video was recorded in Dublin City, Ireland, and depicted the International Financial Services Centre (IFSC), public transport, and people walking through the streets. The video was supported by the sounds of the busy city environment. The 360-degree countryside video was recorded in a rural part of Meath, Ireland and provided a view of a typical Irish countryside setting with bright sunlight, the movement of the grass and the background noises included the sound of the wind.

(a)

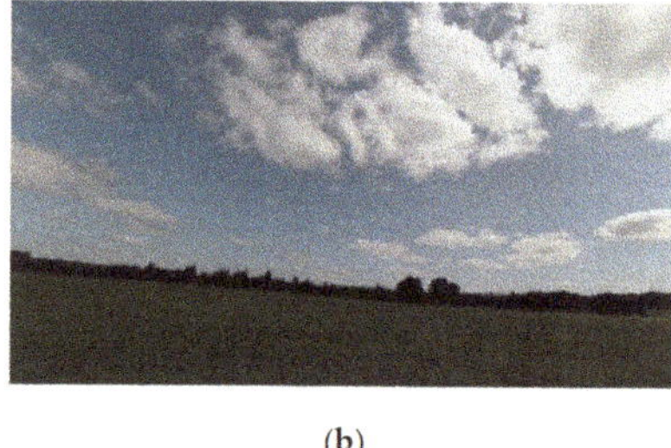

(b)

Figure 2. Contextual settings for the sensory evaluation of chocolate: (**a**) VR busy city (Dublin, Ireland) and (**b**) VR Irish countryside.

For both trials, participants were asked questions regarding their level of engagement, perceived effort to assess the samples, level of distraction, and purchase intent. The following questions were asked for all contextual environments: Was the sensory testing experience memorable? Do you think testing the product in the surrounding environment requires much effort? Do you think the surrounding environment distracted you from performing the task? Responses to these questions were rated on a 7-point scale where

1 = not at all and 7 = very much. Purchase intent of each sample was measured using 5-point scale where 1 = definitely not buy and 5 = definitely buy.

2.4. Data Analysis

Due to the small sample size of participants used in both the beef (n = 30) and chocolate (n = 30) trial, and because the data collected was ordinal, non-parametric tests were used to analyse the data. For the chocolate trial, the Friedman's ANOVA test was performed to determine the effect of the different environmental conditions on participants' hedonic ratings of chocolate samples. Where Friedman's indicated significant differences were present, a Wilcoxon pairwise comparison (with Bonferroni adjustment) was conducted to identify where the differences occurred. In terms of the beef trial, a Wilcoxon signed-rank test was conducted to test if the surrounding environment influenced hedonic ratings of beef samples. The probability level used for significance between responses was set at $p < 0.05$ and all data were analysed using SPSS 25 statistical software (IBM, Armonk, NY, USA).

3. Results and Discussion

3.1. The Effect of the Surrounding Environment on Participants' Hedonic Ratings of Beef Steaks

Tenderness, juiciness, and flavour are the most important palatability attributes affecting beef eating quality [23,24]. In this trial, participants evaluated beef steak in terms of smell, tenderness, juiciness, flavour, and overall liking in two different contextual settings. Participants were unaware that the sample of beef presented in both settings was taken from the same steak. A Wilcoxon signed-rank test showed that the environmental condition had a significant effect of liking of attributes smell ($Z = -2.514$, $p = 0.012$), tenderness ($Z = -3.094$, $p = 0.002$), juiciness ($Z = -3.129$, $p = 0.002$) and flavour ($Z = -2.982$, $p = 0.003$), with beef consumed in the VR restaurant environment setting receiving higher liking scores, compared to the sensory booths. In addition, overall liking scores were significantly greater ($Z = -2.898$, $p = 0.004$) for beef eaten in the VR context, indicating that the VR restaurant had a positive impact on participants' hedonic ratings of beef steaks. Descriptive statistics for the data is presented in Table 1. To the best of our knowledge, no studies to date have investigated the application of immersive technologies, such as VR, in meat sensory research. Hersleth et al., (2015) [25], showed that evoked meal contexts in the form of written texts and images, affected consumers' responses to both the intrinsic and extrinsic attributes in dry-cured ham, although the strongest effects were observed for the extrinsic ratings. In addition, many studies have demonstrated that a restaurant environment is typically associated with positive emotions [26–28], and that consumers tend to be more critical of beef tenderness in the home than in a restaurant [24]. Taken together, these findings indicate that the surrounding environment can considerably influence consumers' sensory perception of beef. As a result, VR technology could provide the food industry with new opportunities to improve sensory marketing efforts through the creation of immersive, multisensory dining experiences, specifically designed to enhance consumer engagement.

The potential for VR in creating a unique eating experience has never been more relevant, as the current coronavirus pandemic continues to hamper access to sharing meal experiences with family and friends. In addition, recent advancements in virtual reality and computational gastronomy could be used an as innovative tool to create multisensory meat-eating experiences while potentially alleviating the health concerns consumers often associate with red meat [13,29]. A more realistic insight into the environmental or contextual cues which influence sensory perception of meat, may also assist industry in developing innovative strategies that will assist in improving the meat-eating experience, and enhance brand characteristics by eliciting stronger emotional connections towards meat products.

Table 1. Means, standard deviations (SD), medians and interquartile ranges (IQR) for participant hedonic ratings * of beef in a traditional sensory booth and a virtual reality (VR) restaurant environment.

Context/Attribute	Sensory Booth				VR Restaurant			
	Mean	SD	Median	IQR	Mean	SD	Median	IQR
Smell	6.67	1.12	6.5	1.25	7.20	0.85	7	1
Tenderness	6.17	2.02	7	3	7.53	1.48	7	2
Juiciness	5.87	1.85	6	2.25	7.03	1.38	7	2
Flavour	5.93	1.62	6	2	6.87	1.41	7	2
Overall liking	6.13	1.85	7	2.25	7.17	1.12	7	1.25

* Evaluated on a 9-point hedonic scale, where 1 = dislike extremely and 9 = like extremely.

3.2. The Effect of the Surrounding Environment on Participants' Hedonic Ratings of Chocolate Samples

In this study, participants rated the smell, flavour, sweetness, texture, smoothness, and overall liking of three identical pieces of chocolate in three different contextual settings. Participants were unaware that the same chocolate was being served in all three conditions. There was a significant difference in perceived liking of flavour ($\chi^2(2) = 7.971$, $p = 0.019$) and overall liking ($\chi^2(2) = 10.750$, $p = 0.005$), depending on the environment in which the chocolate was consumed. A post hoc analysis with Wilcoxon signed-rank tests was conducted with a Bonferroni correction applied, resulting in a significance level set at $p < 0.017$. The results showed that participants rated the flavour ($Z = -2.950$, $p = 0.003$) and overall liking ($Z = -2.885$, $p = 0.004$) of chocolate significantly higher when consumed in the VR countryside setting, compared to the traditional sensory booth. The surrounding context did not influence hedonic ratings of other chocolate attributes. However, in general, the VR countryside setting produced higher liking scores for chocolate compared to the VR busy city and sensory booth environment. Descriptive statistics for the data is presented in Table 2.

Table 2. Means, standard deviations (SD), medians and interquartile ranges (IQR) for participant hedonic rating * of chocolate in a traditional sensory booth, a VR busy city and a VR Irish countryside environment.

Context/Attribute	Sensory Booth				VR Busy City				VR Irish Countryside			
	Mean	SD	Median	IQR	Mean	SD	Median	IQR	Mean	SD	Median	IQR
Smell	6.53	1.78	7	3	6.77	1.46	7	2	7.10	1.35	7	2
Flavour	6.83	1.62	7	2	7.13	1.41	7	2	7.33	1.24	7.5	1.25
Sweetness	7.37	1.22	7	1.5	7.23	1.41	7	1.25	7.13	1.43	7	2.25
Texture	7.03	1.65	7	2.25	6.97	1.25	7	1	7.27	1.23	7	2
Smoothness	7.10	2.04	8	3	7.27	0.98	7	1.25	7.43	1.17	8	1
Overall liking	6.80	1.58	7	2	7.23	1.33	7	2	7.50	1.11	7	1

* Evaluated on a 9-point hedonic scale, where 1 = dislike extremely and 9 = like extremely.

While our research findings agree with previous studies which show that consumers' hedonic ratings of food can change depending on the specific context [2,12,15,25,26], it is also contradictory with others [8,10,11]. For instance, Kong et al. (2020) [11] studied the impact of three contextual settings, including sensory booths and two VR environments (a pleasant sightseeing tour and a live music concert), on consumer liking of three types of chocolate. While their research showed significant differences in consumer liking of the different chocolate types, the surrounding context did not influence the tasting experience. The different immersive environments used to evoke context might explain the inconsistency in findings between this study and ours. Research suggests that certain food products are better suited to specific consumption contexts [3,30], which in turn may alter the sensory perception of food during consumption, through the elicitation of a positive emotional response [26]. Although in this exploratory study we did not measure consumer emotional response to chocolate in the different immersive environments, the more naturalistic and peaceful setting of the Irish countryside might have improved the mood of the participants, which favourably influenced their liking of the chocolate flavour, when compared to the sensory booths and VR city environment [31]. Based on this assumption, we would have expected the chocolate consumed in the busy VR city environment to have produced overall lower hedonic ratings in comparison to the quiet sensory booth setting, as observed elsewhere for chocolate ice-cream [26]. Picket and Dando [12], used VR to investigate how the surrounding environment (winery vs bar) influenced sensory perception of two alcoholic drinks (wine and beer). They showed that while participants liked the wine more and were willing to pay more for it when it was consumed in a virtual winery context, these effects were not observed for beer.

VR technology is a promising tool for simulating immersive environments for consumer sensory evaluations and could be used in conjunction with non-invasive biometric sensors to enrich our understanding of the emotions responsible for driving consumer liking in the real world [2,13,32]. Nonetheless, as research in this area is in its infancy, further studies are required to identify the specific conditions in which utilizing VR is relevant and is likely to improve the reliability and ecological validity of sensory testing outcomes [2,13].

Beyond the importance of selecting an appropriate immersive environment for a particular food product, another possible explanation for the conflicting results associated with using VR as a context-enhancing technology for sensory assessments could be related to how it is being applied during sensory tests. For instance, some studies, including ours, have created immersive VR environments by displaying custom-recorded 360-degree VR videos and their corresponding sounds through a HMD [11,12,15]. In contrast, computer-simulated 3D virtual environments, in which participants can physically walk around and interact with the virtual space using a HMD, have been used to study various aspects of consumer behaviour [33,34]. While both methods are expected to provide a more engaging experience compared to a traditional testing environment [2,5], the use of the HMD fully replaces the participants' view of the real world, restricting their ability to interact visually with the 'real' food product. In an effort to capture consumers' hedonic responses while fully immersed in the VR environment, the participants in this study provided answers to the questions verbally, which were subsequently recorded by the researcher. In similar studies, participants were instructed to remove their VR headsets after tasting and answer questions on a paper ballot [10,11], while othersoverlaid audio instructions and visual scales in the 360-degree VR video [15]. While each approach has clear advantages and disadvantages, future studies could investigate which approach can best improve the ecological validity of consumer sensory data, while continuing to retain participant engagement. In the meantime, this limitation could be overcome by using a mixed-reality device, such as Microsoft HoloLens, which integrates specific elements from the 'real-world' within the surrounding VR space, enabling the user to interact with the real food product while simultaneously answering questions in a controlled environment [13]. Mixed reality was recently utilised to understand consumer response to tea-break snacks and was shown to evoke ecologically valid data comparable to a real-life context [14].

3.3. Engagement, Effort, Distraction and Purchase Intent for Beef Steaks and Chocolate

The mean scores and standard deviations for beef and chocolate are shown in Figure 3a,b, respectively. There were no significant differences in the effort required to perform the sensory test in the respective environments for beef ($p = 0.865$) or chocolate ($p = 0.153$). For beef, a significant difference ($Z = -2.262$, $p = 0.013$) was found in terms of distraction between the two environments, with participants recording higher levels of distraction when immersed in the VR restaurant setting than the sensory booths. This effect was not observed in the chocolate study, although the effect was approaching significance ($p = 0.055$), with scores indicating that the VR busy city environment was more distracting than the sensory booth. Two separate studies by the same research group found contrasting results with respect to the impact of immersion on levels of distraction [2,5]. However, in these studies, the immersive context was presented to participants by means of a purpose-built physical environment, as opposed to a VR headset as used in ours. Nonetheless, efforts to minimise potential distractions contained within an immersive VR environment is required to improve the value and reliability of data generated using this method.

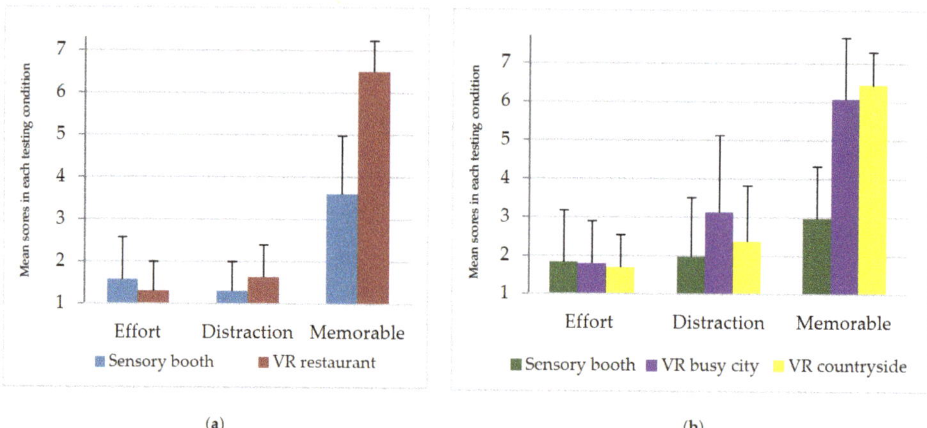

Figure 3. Participants mean scores and standard deviations for beef (**a**) and chocolate (**b**) in each testing condition. The questions asked were: Do you think testing the product in the surrounding environment requires much effort? Do you think the surrounding environment distracted you from performing the task? Was the sensory testing experience memorable? Responses recorded on a 7-point scale where 1 = not at all and 7 = very much.

In our study, participants found the VR restaurant experience to be a significantly more memorable testing environment in comparison to evaluating the beef steaks in the traditional sensory booth set-up ($Z = -4.323$, $p < 0.0001$). A similar trend was observed for the chocolate samples tested in the three different environments, with the VR settings perceived as more memorable compared to the traditional sensory booth ($\chi^2(2) = 46.444$, $p < 0.0001$). In terms of purchase intent, no significant differences were observed between the three environmental conditions for chocolate ($p = 0.124$). For beef, participants were more willing to purchase the beef when consumed in the VR restaurant than in the sensory booth ($p < 0.05$).

3.4. Limitations

While our data yielded some interesting findings with respect to how different VR environments impact hedonic response to beef and chocolate, our study has some limitations that offer avenues for further research. An obvious limitation is the relatively small sample size of participants used in both experimental trials, and therefore further research is needed in a more representative sample of the population. Another potential limitation was that participant hedonic responses were not also collected in a 'real-world' environment, which was not possible in this research study due to economic and time constraints. Finally, it must be noted that improvements are needed in how data are collected from participants immersed in a VR environment, such as including scales in the VR video, or utilizing a mixed-reality approach, in which elements of the 'real-world' environment are integrated within the VR space.

4. Conclusions

This study demonstrated that changing the surrounding environment had a significant effect on participants' hedonic ratings of beef and chocolate. Specifically, we showed that an immersive VR environment tended to induce a positive hedonic response, when compared to a traditional sensory booth setting. Beef was rated as significantly higher for all sensory attributes when consumed in the VR restaurant. For chocolate, the VR countryside context generated significantly higher hedonic scores for flavour and overall liking in comparison to the sensory booth. In general, the VR countryside setting produced the highest liking scores for chocolate. This study also demonstrated that the perceived effort to assess food samples was not significantly impacted by using VR technology, although the surrounding VR environment was more distracting, in comparison to the sensory booths. Overall, this research enhances our understanding of how different environments, simulated using VR technology, can impact sensory perception of beef and chocolate. The study also demonstrates how VR technology can be used to evoke contextual information in consumer sensory tests, which is often an expensive and time-consuming task to conduct in a 'real-world' environment. Nonetheless, as studies investigating the application of VR in stimulating the human senses are only beginning to emerge, further research is necessary before the technology can be confidently used to improve the predictive validity of consumer sensory assessments.

Author Contributions: Conceptualization, E.C. and C.B.; Data curation, C.B.; Formal analysis, E.C. and C.B.; Funding acquisition, E.C., N.M. and C.B.; Investigation, C.B.; Methodology E.C. and C.B.; Project administration, E.C. and C.B.; Resources, E.C. and C.B.; Software, C.B.; Writing—original draft, E.C. and C.B.; Writing—review & editing, E.C., N.M. and C.B. All authors have read and agreed to the published version of the manuscript.

Funding: The authors acknowledge the contribution of Meat Technology Ireland (MTI) a co-funded industry/Enterprise Ireland project (TC 2016 002). The authors also received internal funding from Teagasc.

Institutional Review Board Statement: The sensory trials were conducted in accordance with the guidelines for ethical and professional practices for the sensory analysis of foods as set out by the Institute of Food Science and Technology and the American Meat Science Association.

Informed Consent Statement: Informed consent was obtained from all participants involved in the study.

Data Availability Statement: The data presented in this study is available on request from the corresponding author.

Acknowledgments: Authors would like to thank Cormac McElhinney and Michael Whelan and for their assistance, to Cristina David, Jennifer Zeng and Chris Ovenden for their help with running the trials, to all the staff and students at Teagasc Food Research Centre who participated in the study, and to Daniel Ekhlas for preparing the graphical abstract.

Conflicts of Interest: The authors declare no conflict of interest.

References

1. Jaeger, S.R.; Hort, J.; Porcherot, C.; Ares, G.; Pecore, S.; MacFie, H.J.H. Future directions in sensory and consumer science: Four perspectives and audience voting. *Food Qual. Prefer.* **2017**, *56*, 301–309. [CrossRef]
2. Bangcuyo, R.G.; Smith, K.J.; Zumach, J.L.; Pierce, A.M.; Guttman, G.A.; Simons, C.T. The use of immersive technologies to improve consumer testing: The role of ecological validity, context and engagement in evaluating coffee. *Food Qual. Prefer.* **2015**, *41*, 84–95. [CrossRef]
3. Piqueras-Fiszman, B.; Spence, C. *The Perfect Meal: The Multisensory Science of Food and Dining*; Wiley-Blackwell: Oxford, UK, 2014.
4. Nijman, M.; James, S.; Dehrmann, F.; Smart, K.; Ford, R.; Hort, J. The effect of consumption context on consumer hedonics, emotional response and beer choice. *Food Qual. Prefer.* **2019**, *74*, 59–71. [CrossRef]
5. Hathaway, D.; Simons, C.T. The impact of multiple immersion levels on data quality and panelist engagement for the evaluation of cookies under a preparation-based scenario. *Food Qual. Prefer.* **2017**, *57*, 114–125. [CrossRef]
6. Köster, E.P. The psychology of food choice: Some often encountered fallacies. *Food Qual. Prefer.* **2003**, *14*, 359–373. [CrossRef]
7. Hein, K.A.; Hamid, N.; Jaeger, S.A.; Delahunty, C.M. Application of a written scenario to evoke a consumption context in a laboratory setting: Effects on hedonic ratings. *Food Qual. Prefer.* **2010**, *21*, 410–416. [CrossRef]
8. van Bergen, G.; Zandstra, E.H.; Kaneko, D.; Dijksterhuis, G.B.; de Wijk, R.A. Sushi at the beach: Effects of congruent and incongruent immersive contexts on food evaluations. *Food Qual. Prefer.* **2021**, *91*, 104193. [CrossRef]
9. Zandstra, E.H.; Daneko, D.; Dijksterhuis, G.B.; Vennik, E.; de Wijk, R.A. Implementing immersive technologies in consumer testing: Liking and Just-About-Right ratings in a laboratory, immersive simulated café and real café. *Food Qual. Prefer.* **2020**, 103934. [CrossRef]

10. Torrico, D.D.; Han, Y.; Sharma, C.; Fuentes, S.; Gonzalez Viejo, C.; Dunshea, F.R. Effects of context and virtual reality environments on the wine tasting experience, acceptability, and emotional responses of consumers. *Foods* **2020**, *9*, 191. [CrossRef]
11. Kong, Y.; Sharma, C.; Kanala, M.; Thakur, M.; Li, L.; Xu, D.; Harrison, R.; Torrico, D. Virtual reality and immersive environments on sensory perception of chocolate products: A preliminary study. *Foods* **2020**, *9*, 515. [CrossRef]
12. Picket, B.; Dando, R. Environmental immersion's influence on hedonics, perceived appropriateness, and willingness to pay in alcoholic beverages. *Foods* **2019**, *8*, 42. [CrossRef] [PubMed]
13. Crofton, E.C.; Botinestean, C.; Fenelon, M.; Gallagher, E. Potential applications for virtual and augmented reality technologies in sensory science. *Innov. Food Sci. Emerg. Technol.* **2019**, *56*, 102178. [CrossRef]
14. Low, J.Y.Q.; Lin, V.H.F.; Jun Yeon, L.; Hort, J. Considering the application of a mixed reality context and consumer segmentation when evaluating emotional response to tea break snacks. *Food Qual. Prefer.* **2021**, *88*, 104113. [CrossRef]
15. Stelick, A.; Penano, A.G.; Riak, A.C.; Dando, R. Dynamic context sensory testing–A proof of concept study bringing virtual reality to the sensory booth. *J. Food Sci.* **2018**, *83*, 2047–2051. [CrossRef] [PubMed]
16. Kiefer. How Guinness and R/GA Made a VR Tasting Experience for all Five Senses. 2017. Available online: https://www.campaignlive.co.uk/article/guinness-r-ga-made-vr-tasting-experience-five-senses/1434731 (accessed on 25 March 2021).
17. Brinkley. I Ate a Meal in Virtual Reality. Here's What it Tasted Like. 2020. Available online: https://www.cnbc.com/2020/03/21/virtual-reality-dining-explained.html (accessed on 25 March 2021).
18. Seo, H.S. Sensory nudges: The influences of environmental contexts on consumers' sensory perception, emotional responses, and behaviors toward foods and beverages. *Foods* **2020**, *9*, 509. [CrossRef]
19. Elzerman, J.; Hoek, A.C.; van Boekel, M.A.J.S.; Luning, P.A. Consumer acceptance and appropriateness of meat substitutes in a meal context. *Food Qual. Prefer.* **2011**, *22*, 233–240. [CrossRef]
20. Piqueras-Fiszman, B.; Jaeger, S. The impact of evoked consumption contexts and appropriateness on emotion responses. *Food Qual. Prefer.* **2014**, *32*, 277–288. [CrossRef]
21. Institute of Food Science and Technology. IFST Guidelines for Ethical and Professional Practices for the Sensory Analysis of Foods. 2015. Available online: http://www.ifst.org/knowledge-centre-other-knowledge/ifst-guidelines-ethical-and-professional-practices-sensory (accessed on 20 January 2020).
22. AMSA. *Research Guidelines for Cookery, Sensory Evaluation and Instrumental Tenderness Measurements of Meat*, 2nd ed.; American Meat Science Association in Cooperation with National Live Stock and Meat Board: Champaign, IL, USA, 2015.
23. Botinestean, C.; Hossain, M.; Mullen, A.M.; Kerry, J.P.; Hamill, R.M. The influence of the interaction of sous-vide cooking time and papain concentration on tenderness and technological characteristics of meat products. *Meat Sci.* **2021**, *177*, 108491. [CrossRef]
24. Miller, M.F.; Hoover, L.C.; Cook, K.D.; Guerra, A.L.; Huffman, K.L.; Tinney, K.S.; Ramsey, C.B.; Brittin, H.C.; Huffman, L.M. Consumer acceptability of beef steak tenderness in the home and restaurant. *J. Food Sci.* **1995**, *60*, 963–965. [CrossRef]
25. Hersleth, M.; Monteleone, E.; Segtnan, A.; Naes, T. Effects of evoked meal contexts on consumers' responses to intrinsic and extrinsic product attributes in dry-cured ham. *Food Qual. Prefer.* **2015**, *40*, 191–198. [CrossRef]
26. Xu, Y.; Hamid, N.; Shepherd, D.; Kantono, K.; Spence, C. Changes in flavour, emotion, and electrophysiological measurements when consuming chocolate ice cream in different eating environments. *Food Qual. Prefer.* **2019**, *77*, 191–205. [CrossRef]
27. Danner, L.; Ristic, R.; Johnson, T.E.; Meiselman, H.L.; Hoek, A.C.; Jeffery, D.W.; Bastian, S.E. Context and wine quality effects on consumers' mood, emotions, liking and willingness to pay for Australian Shiraz wines. *Food Res. Int.* **2016**, *89*, 254–265. [CrossRef] [PubMed]
28. Prayag, G.; Khoo-Lattimore, C.; Sitruk, J. Casual dining on the French Riviera: Examining the relationship between visitors' perceived quality, positive emotions, and behavioral intentions. *J. Hosp. Mark. Manag.* **2014**, *24*, 24–46. [CrossRef]
29. Velasco, C.; Obrist, M.; Petit, O.; Spence, C. Multisensory technology for flavour augmentation: A mini review. *Front. Psychol.* **2018**, *9–26*. [CrossRef]
30. Cardello, A.V.; Pineau, B.; Paisley, A.G.; Roigard, C.M.; Chheang, S.L.; Guo, L.F.; Hedderley, H.; Jaeger, S.R. Cognitive and emotional differentiators for beer: An exploratory study focusing on "uniqueness". *Food Qual. Prefer.* **2016**, *54*, 23–38. [CrossRef]
31. Boutrolle, I.; Delarue, J.; Arranz, D.; Rogeaux, M.; Köster, E.P. Central location test vs. home use test: Contrasting results depending on product type. *Food Qual. Prefer.* **2007**, *18*, 490–499. [CrossRef]
32. Viejo, C.G.; Villarreal-Lara, R.; Torrico, D.D.; Rodríguez-Velazco, Y.G.; Escobedo-Avellaneda, Z.; Ramos-Parra, P.A.; Mandal, R.; Singh, A.P.; Hernández-Brenes, C.; Fuentes, S. Beer and Consumer Response Using Biometrics: Associations Assessment of Beer Compounds and Elicited Emotions. *Foods* **2020**, *9*, 821. [CrossRef] [PubMed]
33. Siegrist, M.; Ung, C.Y.; Zank, M.; Marinello, M.; Kunz, A.; Hartmann, C.; Menozzi, M. Consumers' food selection behaviors in three-dimensional (3D) virtual reality. *Food Qual. Prefer.* **2019**, *117*, 50–59. [CrossRef]
34. Ung, C.Y.; Menozzi, M.; Hartmann, C.; Siegrist, M. Innovations in consumer research: The virtual food buffet. *Food Qual. Prefer.* **2018**, *63*, 12–17. [CrossRef]

 *foods*

Article

Virtual Reality and Immersive Environments on Sensory Perception of Chocolate Products: A Preliminary Study

Yanzhuo Kong, **Chetan Sharma**, **Madhuri Kanala**, **Mishika Thakur**, **Lu Li**, **Dayao Xu**, **Roland Harrison** and **Damir D. Torrico** *

Department of Wine, Food and Molecular Biosciences, Lincoln University, Lincoln 7647, New Zealand;
yanzhuokong@outlook.com (Y.K.); chetan.sharma@lincoln.ac.nz (C.S.);
Madhuri.KanalaManjunath@lincolnuni.ac.nz (M.K.); Mishika.Thakur@lincolnuni.ac.nz (M.T.);
Lu.Li2@lincolnuni.ac.nz (L.L.); Dayao.Xu@lincolnuni.ac.nz (D.X.); Roland.Harrison@lincoln.ac.nz (R.H.)
* Correspondence: Damir.Torrico@lincoln.ac.nz; Tel.: +64-3-423-0641

Received: 23 March 2020; Accepted: 16 April 2020; Published: 20 April 2020

Abstract: Traditional booths where sensory evaluation usually takes place are highly controlled and therefore have limited ecological validity. Since virtual reality (VR) is substantially interactive and engaging, it has the potential to be applied in sensory science. In this preliminary study, three chocolate types (milk, white, and dark) were evaluated under three contextual settings, including sensory booths (control) and two VR environments (360-degree videos using VR headsets: (i) a pleasant sightseeing tour, and (ii) a live music concert). Untrained participants ($n = 67$) were asked to rate their liking and the intensity of different chocolate attributes based on the 9-point hedonic scale and just-about-right-scale (JAR). Emotions were evaluated using the check-all-that-apply (CATA) method. Results showed that there were no significant effects of context type on the tasting experience; however, there were significant effects of chocolate type. Milk and white chocolates were preferred over dark chocolate irrespective of the context type. Additionally, more positive emotions were elicited for the dark chocolate in the "virtual live concert" environment. Dark chocolate under the other two environments was associated with negative emotional terms, such as "bored" and "worried." In terms of more reliable and ecologically valid sensory responses, further research is needed to match suitable VR environments to different chocolate types.

Keywords: virtual reality; immersive environments; acceptability; emotions; chocolate products

1. Introduction

Virtual reality (VR) is an emerging technology that provides artificially simulated environments based on computer technology and relevant software [1]. These artificial environments can be either recorded videos, pictures, or animated scenes, which are either immersive [2] or non-immersive [3], and either similar or completely different from the real world [4]. In general, both immersive and non-immersive VR environments can be achieved by some common methods, such as the wall projection [2]. However, fully immersive VR environments are usually achieved by VR headsets nowadays. Based on the high-resolution 360-degree vision and 3D sound, VR headsets provide users with a highly interactive and engaging experience by mobilizing their sight and hearing [5].

VR has become increasingly common in our daily life. For example, VR can enrich people's entertainment life on the basis of immersive sensory experience regarding games, movies, travelling, and even shopping [6]. It has also been commonly applied in training and education areas, such as supporting the teaching and learning process [7]. A few studies have also explored its application in medicine and tourism [8,9]. Apart from that, applying VR technologies in sensory science is an emerging research field due to the limited ecological validity of traditional sensory booths [10].

The eating environment is considered to have a significant impact on consumers' sensory perception as well as hedonic responses towards most food products [10]. In sensory science, booths can be used for the sensory evaluation of food products. However, the sensory responses obtained under booths may not describe the totality of the eating experience since the testing environment is highly controlled [10]. This could be one of the reasons why many new products, which were the most liked in consumer trials, eventually failed after launching into the marketplace [11]. The highly controlled sensory booths do not reflect the real consumption environment, and eventually, the obtained results from such setups could have a low ecological validity [12]. However, it seems challenging and relatively expensive to evaluate food products in various real environments. Therefore, the interest in using VR technologies in sensory science has increased dramatically in recent years.

The application of VR technologies in sensory science helps us better understand ecologically valid consumer experiences for certain food products [13]. A few studies have applied VR technologies in sensory science by simulating physically immersive environments, such as a bar, a coffeehouse, and an airplane [14–16]. In general, the perceived appropriateness and enjoyment of the eating process for certain food products could be largely influenced by the context. Sinesio et al. [2] observed higher liking scores of tomato and wild rocket salads when they were tasted in an immersive multisensory room compared to the traditional sensory booths. However, the discrimination efficacy of freshness of both vegetables was lower in the immersive environment than in the booths. As different food categories could elicit different sensory perceptions from participants, it is also necessary to take food categories into consideration. For example, Picket and Dando [17] tested how context influences the sensory perception of two alcoholic drinks, beer and sparkling wine. The results showed that the suitable and matched environments for each alcoholic drink tended to improve participants' acceptance and willingness to purchase.

As a common snack, chocolate tends to elicit more emotions than other snacks, such as chips [18]. Therefore, three major chocolate products, which were white chocolate, milk chocolate, and dark chocolate, have been evaluated in this preliminary research to better understand consumers' sensory acceptability and emotional responses affected by immersive VR environments. Three contextual settings were applied in the sensory evaluation process, including two 360-degree recorded videos using VR headsets, and traditional sensory booths as the control setting. As there is an increasing awareness of limited ecological validity in sensory tests, this study also aimed to preliminarily explore the potential of VR technology as a support in regular sensory tests.

2. Materials and Methods

2.1. Participants

A total of 67 untrained participants (31 males and 36 females, 20–50 years old, originally from Asia, South America, Oceania, and Europe) were recruited voluntarily through Lincoln University intranet. All participants claimed that they were not allergic to the chocolate ingredients involved in this research. A brief introduction about products and sensory procedures was given first. According to Lincoln University Policies and Procedures, they were asked to complete the consent form regarding human ethics (approval: 2019-68) before tasting the chocolate products. Sensory sessions were carried out at the sensory laboratory located in the Replacement For Hilgendorf (RFH) building, Lincoln University, Lincoln, New Zealand. There were three sessions conducted on four consecutive days: one control session using sensory booths and two sessions using VR settings. The duration of each session was around 15 to 25 min per participant, and the order of the three sessions was randomized for each participant. Participants were asked to refrain from eating, drinking, and smoking for at least one hour before the sessions.

2.2. Stimuli

Three major chocolate types were used in this research, including Whittaker's 28% Cocoa White Chocolate, Whittaker's 33% Cocoa Creamy Milk Chocolate, and Whittaker's 72% Cocoa Dark Ghana Chocolate (J.H. Whittaker and Sons, Ltd., Porirua, New Zealand). These chocolate products were purchased from a local supermarket before conducting sensory sessions. They were purchased in blocks and served in squares, and they were stored in sealed containers at 4 °C in a refrigerator (Samsung, Seoul, South Korea) when they were not in use. The preparation and sampling processes were conducted within two hours prior to the sensory sessions to prevent chocolate samples from being stale. Three stimuli (white, milk, and dark chocolate) were assessed initially by a focus group panel ($n = 4$) within Lincoln university to make sure they had notable differences in terms of certain attributes, such as sweetness and cocoa flavor. In each sensory session, each chocolate square was served in a transparent plastic cup coded with a 3-digit random number for identification. The presentation order of the three samples was randomized and balanced for each participant to prevent positional bias.

2.3. Sensory Procedure

At the beginning of sensory sessions, a brief explanation of the procedures was given to all participants. They were instructed regarding the proper operation and wearing of the VR headsets as well as how to answer questions in tablets (Galaxy Tab A, Samsung, Seoul, South Korea). Three sensory sessions were randomly carried out for each participant. After signing the consent form, participants were instructed to evaluate three randomly ordered chocolate samples from left to right under one of three contextual settings (booth or two VR settings). In the questionnaire, participants were asked to rate the acceptability of ten attributes of each chocolate sample, including taste/flavor, sweetness, bitterness, cocoa flavor, dairy flavor, texture, hardness, smoothness, aftertaste, and overall liking. The acceptability test was based on the 9-point hedonic scale, which represented nine hedonic responses using number 1 (dislike extremely) to 9 (like extremely) with a neutral response at 5 (neither like nor dislike) [19]. Sweetness, bitterness, cocoa flavor, dairy flavor, and overall texture were also evaluated by a just-about-right-scale (JAR) in terms of both intensity and acceptability (1 = too little, 2 = just about right, 3 = too much for sweetness, bitterness, cocoa flavor, and dairy flavor; 1 = too soft, 2 = just about right, 3 = too hard for overall texture) [20].The next question was purchase intent of each chocolate sample (Would you purchase this product if it was available at a reasonable price where you normally shop?), and the answer was based on a binomial scale (1 = No, 2 = Yes). The last section of this questionnaire was the evaluation of emotional responses (Please select all the emotions that you think apply regarding this chocolate sample). The check-all-that-apply (CATA) method was used with a list of 33 emotional terms that were pre-selected previously from 48 original terms [21,22]. These terms were "adventurous," "satisfied," "active," "affectionate," "calm," "energetic," "enthusiastic," "free," "friendly," "glad," "good," "happy," "interested," "joyful," "loving," "merry," "nostalgic," "peaceful," "pleased," "pleasant," "secure," "warm," "bored," "disgusted," "worried," "aggressive," "daring," "eager," "guilty," "polite," "steady," "understanding" and "wild." Plain crackers and water were used to cleanse participants' palate in between different chocolate samples.

2.4. Contextual Settings

Three contextual settings were used in this research, including the traditional sensory booth as the control setting, and two VR settings achieved by VR headsets (Oculus VR, LLC., Menlo Park, CA, USA). The sensory booth was located at the RFH building, Lincoln University, Lincoln, New Zealand. Individual booth units were separated by a solid protection panel, and there was a worktop within each isolated booth unit for placing samples and tablets (Figure 1a,c). The booth temperature was set to 21 °C, and white-color fluorescent lights were used during the entire sensory evaluation process.

For sensory sessions associated with the two VR settings, they were carried out in an isolated focus room, which differed from the booth (Figure 1b). Four Oculus Go All-in-One VR Headsets (32 GB) with independent controllers were used for the generation of immersive VR environments. Both VR environments were 360-degree videos selected from a video, movie, and photo platform called VeeR VR, which was available within the VR headset itself (VeeR VR, San Francisco, CA, USA). VeeR VR is a premium VR entertainment platform with more than ten thousand high-quality videos, photos, and interactive experiences. The VR settings used for this research were selected by a focus group ($n = 4$) from a list of 14 preliminarily evaluated 360-degree videos (VeeR VR). The criterion for selecting the videos was based on the experienced pleasantness that participants had during the test in the focus group session. One pleasant and one unpleasant video were selected for further analysis. There were at least two instructors to help participants wear the VR headsets and place samples during the entire sensory session under the VR settings. Chocolate samples were served to participants after they were wearing the headsets and earphones. Participants were instructed to take the headsets off and start answering questions on the tablets when they finished the tasting. This process was repeated until the tasting of all three chocolate samples has been completed for each participant under each VR setting.

With regard to the VR environments, two 360-degree recorded videos (VeeR VR) that could elicit different feelings and emotional responses from participants were selected. The first VR environment was titled as "Pure relaxation in the luxurious apartments Vidamar resorts Algarve." As shown in Figure 1d1–d3, this video was a sightseeing tour in a 5-star hotel located in Guia, Portugal. It had a duration of 41 min along with relaxing music, which showed beautiful sceneries such as a large swimming pool and a peaceful beach. The second VR environment was titled "Elemental Live—Halloween" (Figure 1e1,e2), which had a duration of 54 min. It was a noisy live music concert, which was held on a cloudy day and was crowded with people. The first and second VR environments were represented by the labels positive VR (PVR) and negative VR (NVR) respectively for convenience, and B was used to represent traditional sensory booths as well.

Figure 1. Contextual settings for the sensory evaluation of chocolate products. (**a**) Booths set up; (**b**) Virtual reality (VR) set up; (**c**) Sensory booth setting; (**d**, **1–3**) Positive VR setting; (**e**, **1–2**) Negative VR setting. VR environments were obtained from the VeeR VR app (VeeR VR, San Francisco, CA, USA).

2.5. Statistical Analysis

A 3 × 3 factorial design was used in this research, which referred to three chocolate products and three contextual settings. The acceptability data for sensory attributes were tested for normality by the Shapiro–Wilk test procedure and all attributes that were found violating the normal distribution were transformed through the PROC RANK procedure of Statistical Analysis System-SAS (SAS, Cary, NC, USA) for the subsequent Friedman's test (RStudio, Version 1.1.456—© 2009-2018 RStudio, Inc.). To assess whether the reported sensory differences among chocolate type × environment were significant or not, we used the non-parametric Friedman and post-hoc Nemenyi tests. The Friedman test first establishes whether at least one of the treatments is significantly different from the rest, and if this was the case, we used the Nemenyi test to identify those treatments for which there is evidence of statistically significant differences. Nememyi test can be used [23,24], provided that data has an equal sample size between each group and Friedman-type ranking of the data. The advantage of this testing approach is that it does not impose any distributional assumptions and does not require multiple pairwise testing between treatments, which would distort the outcome of the tests. The package *tsutils* for RStudio implications were used to run the Nemenyi test (R package version 0.9.2.) [25]. Penalty analysis was applied to the JAR data in order to determine how much the overall liking and acceptance of the chocolate samples were influenced by their attributes. The effect of contextual settings was also considered in this research. The CATA emotional responses of chocolate samples under different settings were assessed by the Cochran's Q test, Correspondence analysis (CA), and principal coordinate analysis (PCoA) [22,26]. With regard to the purchase intent, it was analyzed for multiple comparisons based on Cochran's Q test and the simultaneous confidence intervals testing as well. Principal component analysis (PCA; Correlation Biplot) was used to analyze the relationship between the hedonic acceptability of ten attributes and chocolate samples under different environmental settings. The PCA results are presented as a product-attribute biplot. Chocolate samples under different settings were categorized by the Agglomerative Hierarchical Cluster (AHC) analysis. The dissimilarity of them was analyzed based on the Euclidean distance and the Ward's method. The responses given through an electronic questionnaire were collected by RedJade Sensory Software (Martinez, CA, USA). Data were analyzed using Minitab 18 (Minitab, LLC, State College, PA, USA) and XLSTAT Statistical Software 2016 (Addinsoft, New York, NY, USA).

3. Results

3.1. The Effect of Environments on Sensory Acceptability of Chocolate Products

3.1.1. Hedonic Ratings

The sensory attributes measured on 9-point interval scale were analyzed under the null hypothesis that all treatment groups were taken from populations with the same median, and they do not evoke different liking in consumers. Friedman's two-way nonparametric analysis of variance (ANOVA) was used to evaluate the effect of context environment and chocolate type on the response. By using the ANOVA procedure in conjunction with the PROC RANK procedure, data was ranked 'not' by block to analyze the effect of context, whereas traditional method of rank assigning was used, i.e., rank within blocks for the chocolate type. As Friedman's test cannot be used for interaction, another method was used, where the non-normal distributions were transformed by NORMAL = BLOM function of PROC RANK [27]. This function computes normal scores from the ranks using the formula $yi = (1/P)$ $(ri\text{-}3/8)/(n + 1/4)$, and after normalization the data were evaluated by using PROC GLM (general linear model) procedure for main and interaction effects. Each response was grouped by two categorical factors, namely context type, and chocolate type, and ranks were assigned by PROC RANK in SAS. In a nutshell, the chocolate type effect was statistically significant, whereas the environment effect was not significant at 95% confidence level. Furthermore, the interaction between chocolate types and environments did not significantly affect the liking scores of the evaluated sensory attributes at

95% confidence. Interestingly, the cocoa flavor was affected by the environment effect at about 88% confidence level.

Consumers that tasted the different chocolate types showed a significant difference in the liking of sensory attributes and thus, accepting the alternate hypothesis that all treatments groups do not belong to the same population median. Nemenyi test, similar to Tukey-HSD (honestly significant difference), helped identify the treatments responsible for the significance of the chocolate effect by comparing the medians of distribution (Figure 2). In general, milk chocolate had the highest liking scores of the evaluated sensory attributes, followed by white chocolate. However, the two types of chocolate products (milk and white) were not significantly different (sharing the same color, brown line) regarding the liking scores of most attributes except for cocoa flavor (Figure 2d). Milk chocolate under PVR and white chocolate under B scored 6.51 ± 0.20 and 5.06 ± 0.24, respectively, for the liking of cocoa flavor, which were significantly different from each other. Dark chocolate was the least liked chocolate type (aqua color line), and its liking scores of taste/flavor, dairy flavor, texture, smoothness, aftertaste, and overall liking were significantly different from other two chocolate types regardless of the environments. Considering the effect of environments, there were no significant differences among PVR, NVR, and B within the same type of chocolate product ($p > 0.05$). The liking scores of the evaluated attributes under PVR were similar and generally high (but not significant, $p > 0.05$) for both milk chocolate and white chocolate, whereas generally high (but not significant, $p > 0.05$) liking scores of evaluated attributes were obtained under NVR for dark chocolate.

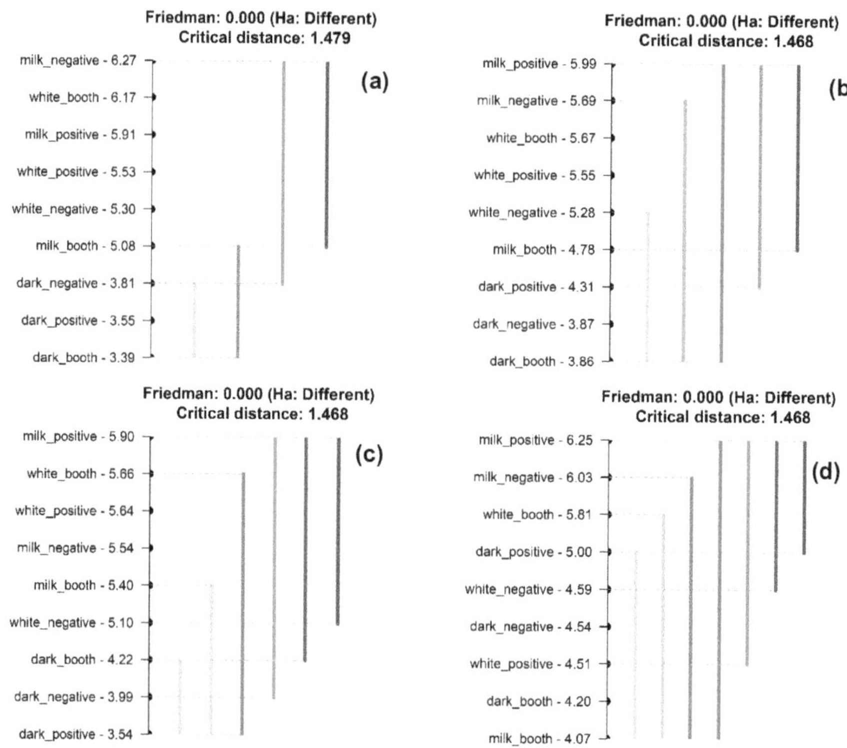

Figure 2. *Cont.*

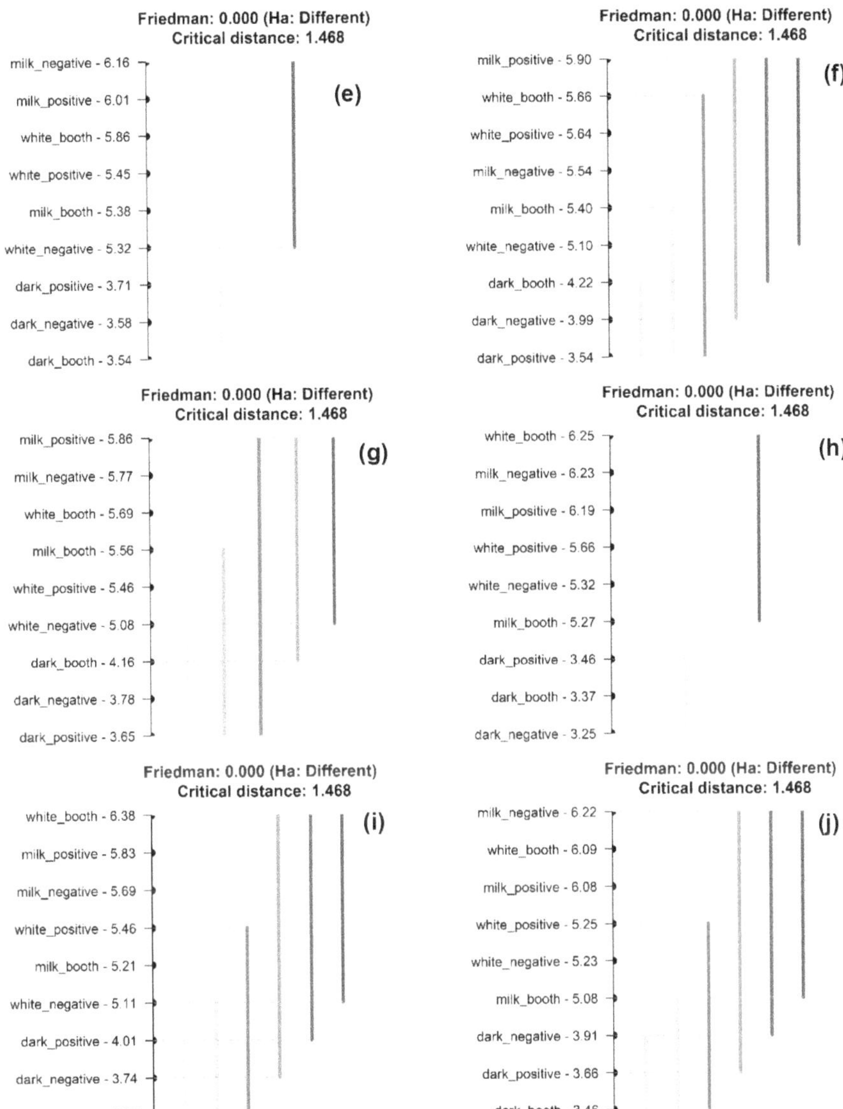

Figure 2. Nemenyi test (post hoc) results comparing medians at 5% significance level for sensory attributes, (**a**) taste, (**b**) sweetness, (**c**) bitterness, (**d**) cocoa flavor, (**e**) dairy flavor, (**f**) texture, (**g**) hardness, (**h**) smoothness, (**i**) aftertaste, and (**j**) overall. Treatments sharing color implies not statistically different.

3.1.2. Just-about-Right (JAR) Results

Figures 3–5 show the JAR frequencies (%) and mean drops based on the penalty analysis for the three chocolate products (milk, white, and dark) under three environments (PVR, NVR, and B), regarding their sweetness, bitterness, cocoa flavor, dairy flavor, and overall texture. As shown in Figure 3, milk chocolate had the highest selections of JAR for cocoa flavor (75%) and overall texture (81%) under B, whereas the highest proportion of participants (75%) selected JAR for dairy flavor under PVR. Milk chocolate under NVR had the highest selections of "too little/soft" for both cocoa flavor (31%) and overall texture (27%), and "too much" for dairy flavor (27%). The frequencies of sweetness and bitterness for milk chocolate were similar under three environments. For white chocolate (Figure 4), JAR was selected most frequently for sweetness (46%) and overall texture (87%) under B, as well as for bitterness (46%) and cocoa flavor (54%) under PVR. White chocolate under NVR had the highest selection of JAR for dairy flavor (63%). Selections of "too much/hard" for bitterness (0–1%), cocoa flavor (0%) and overall texture (4–9%), as well as "too little" for sweetness (0–1%) and dairy flavor (3–10%) of white chocolate were negligible regardless of the environments. With regard to dark chocolate (Figure 5), JAR was selected most frequently for sweetness (46%), bitterness (40%), cocoa flavor (54%), and overall texture (63%) under PVR. Dark chocolate under NVR had the highest selection of JAR for dairy flavor (37%). The frequencies of "too much" regarding sweetness, dairy flavor, and "too little/soft" regarding bitterness, cocoa flavor, the overall texture of dark chocolate under three environments were negligible.

Penalty analysis was conducted based on both JAR frequencies and the overall liking scores of chocolate products considering the environments. As shown in Figures 3–5, the threshold for the population size was set as 20%. The attributes that appeared in the upper right-hand corner of the penalty plot were considered to have negative effects on the liking of products, as more than 20% of people thought they were either "too much/hard" or "too little/soft" [28]. Accordingly, both milk and white chocolate were penalized for being too sweet and not bitter enough, which was opposite to dark chocolate. Milk chocolate under both PVR and NVR was penalized due to not having enough cocoa flavor. In addition, "too much" dairy flavor for milk chocolate under both NVR and B also affected their overall liking scores. The penalty analysis results for both white chocolate and dark chocolate were generally consistent under the three tested environments, of which cocoa flavor and dairy flavor were penalized for being "too little" and "too much" for white chocolate, respectively, whereas the opposite happened with dark chocolate. Moreover, the liking of dark chocolate was affected by its hard texture as well (Figure 5).

3.2. Multivariate Analysis of Chocolate Products under Different Environments

3.2.1. Emotional Responses

According to Cochran's Q test results (Table S1), 23 emotional terms were significant, including "adventurous," "satisfied," "active," "calm," "affectionate," "energetic," "enthusiastic," "friendly," "glad," "good," "happy," "interested," "joyful," "loving," "peaceful," "pleased," "pleasant," "bored," "disgusted," "worried," "aggressive," "polite", and "wild." Figure 6 shows the correspondence analysis (CA) and principal coordinate analysis (PCoA) results. The CA displays the relationships between emotional terms obtained based on the CATA method and three chocolate products considering the contextual effect. As shown in Figure 6a, the principal component one (PC1) and principal component two (PC2) were 63.43% and 22.86%, respectively, which explained 86.29% of data variability in total. According to CA results, milk chocolate and white chocolate under PVR and B were found to share similar profiles, which were associated with both neutral and positive emotional descriptors such as "peaceful," "pleasant," "good," "satisfied," "glad," "pleased," and "polite." Milk and white chocolate under NVR were also associated with positive terms, such as "affectionate," "interested," "happy," "loving," "joyful," and "friendly." Dark chocolate had highly distinctive groups of emotional terms under NVR and PVR/B. With regard to dark chocolate under NVR, it was related to ardent descriptors, including "adventurous," "energetic," "wild," "active," and "enthusiastic." In contrast, dark chocolate under PVR and B were related to negative terms, such as "bored," "worried," "disgusted," and "aggressive."

The PCoA results show the relationship between emotional terms and the overall liking scores of the three chocolate products under three different contextual settings (Figure 6b). Only the terms "aggressive," "disgusted," "worried," and "bored" were selected in relation to the lowest mean values (<5.0) for the overall liking of chocolate products under different environments. However, terms such as "pleased," "glad," "good," "loving," "friendly," "peaceful," "pleasant," "affectionate," "satisfied," and "joyful" contributed to higher overall liking scores of chocolate products considering the contextual effect (>5.0).

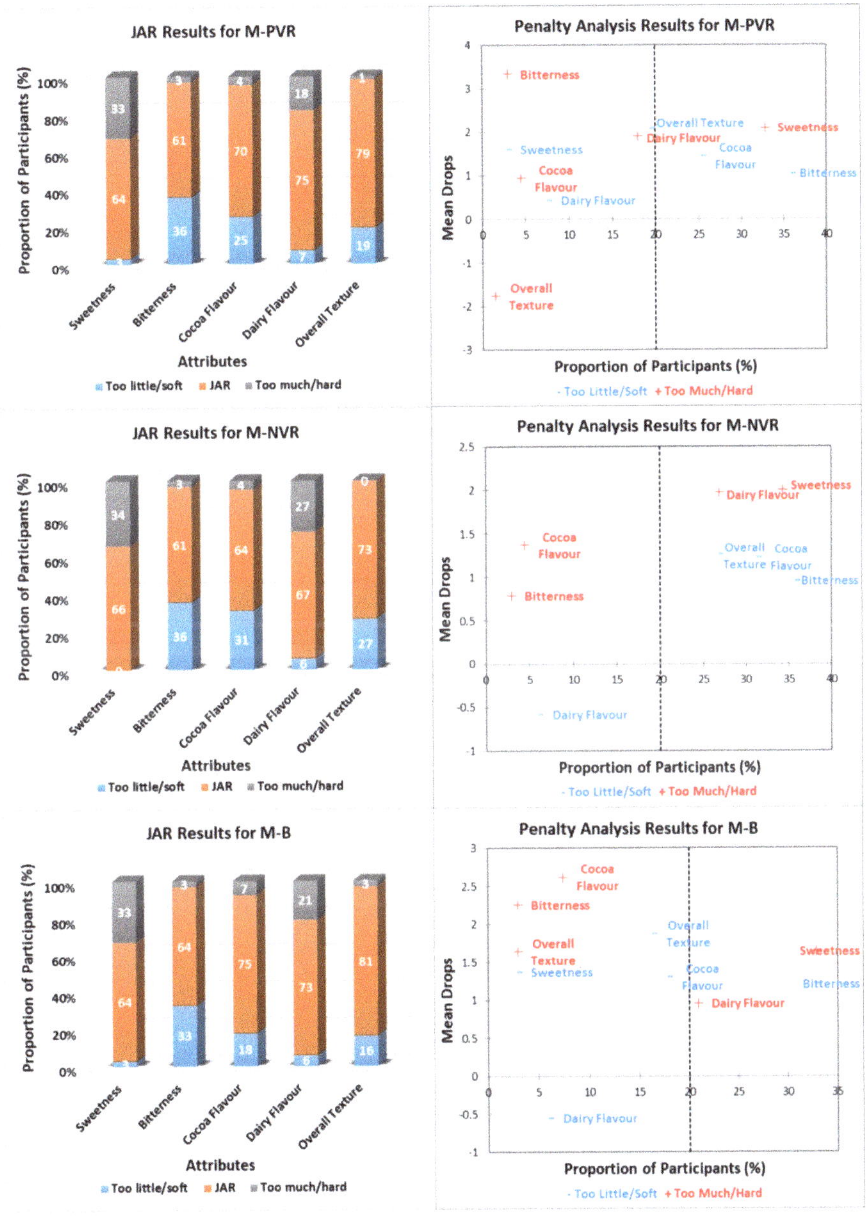

Figure 3. Just-About-Right (JAR) frequencies and penalty analysis results [1] regarding milk chocolate attributes under different environments [2]. [1] Penalty analysis was associated with the overall liking scores (9-point hedonic scale). [2] M-PVR: milk chocolate positive VR; M-NVR: milk chocolate negative VR; M-B: milk chocolate sensory booth.

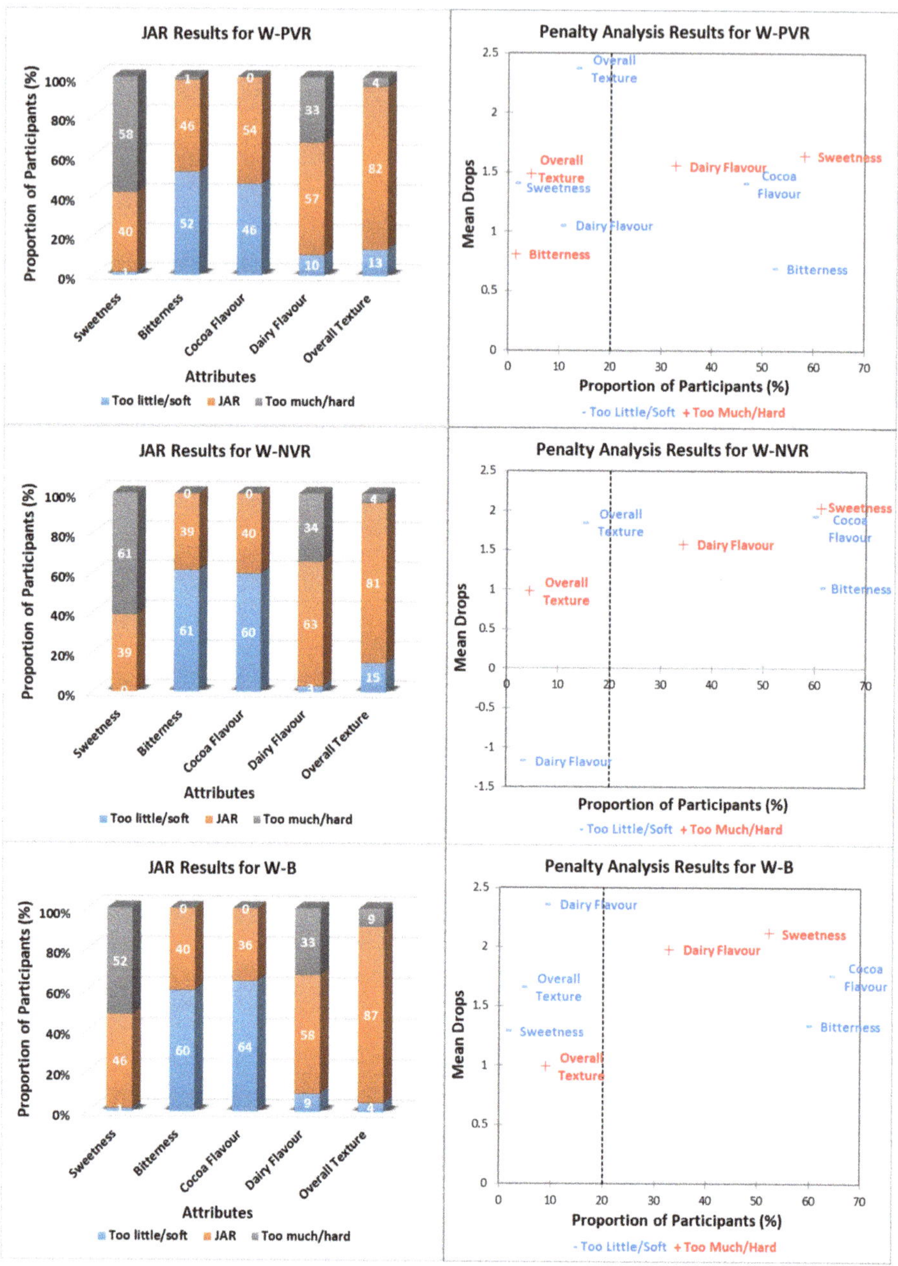

Figure 4. Just-About-Right (JAR) frequencies and penalty analysis results [1] regarding white chocolate attributes under different environments [2]. [1] Penalty analysis was associated with the overall liking scores (9-point hedonic scale). [2] W-PVR: white chocolate positive VR; W-NVR: white chocolate negative VR; W-B: white chocolate sensory booth.

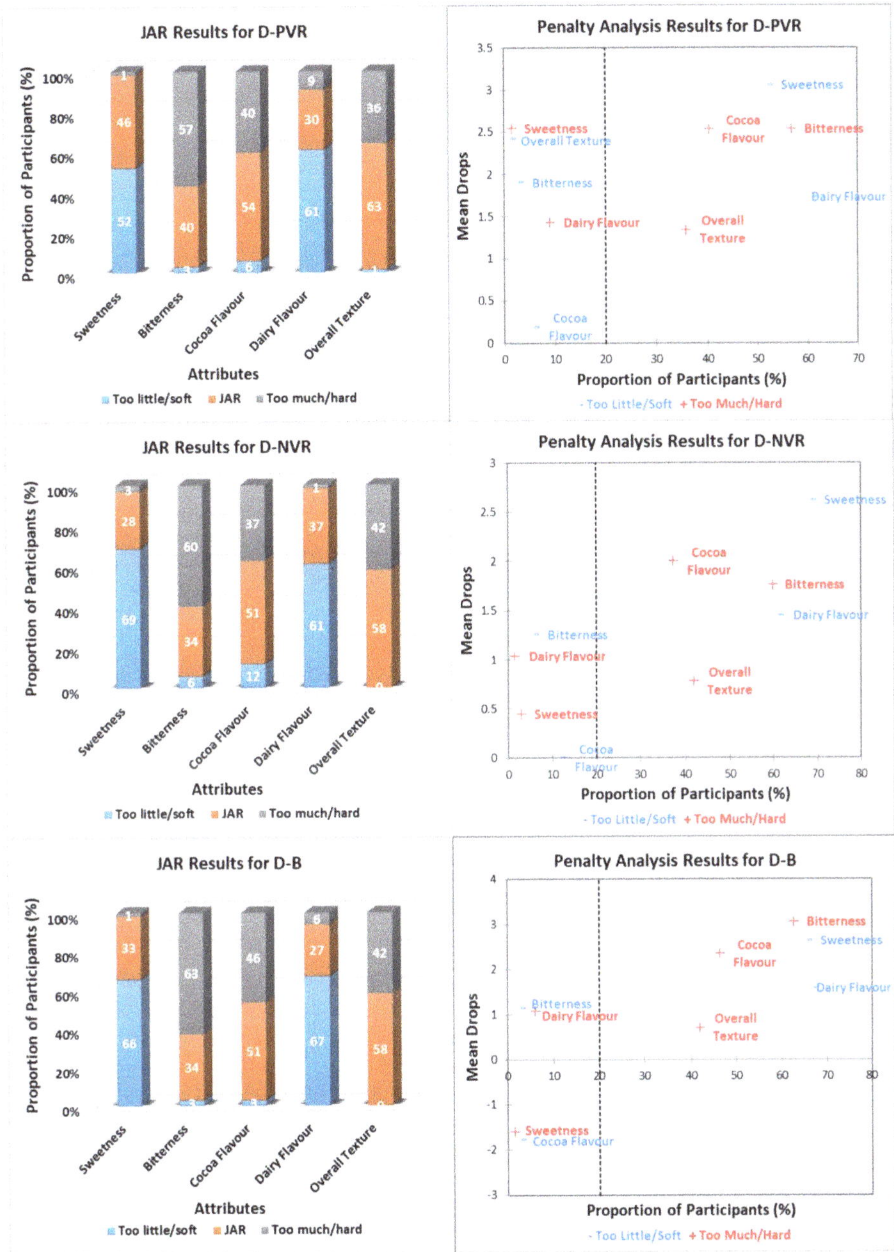

Figure 5. Just-About-Right (JAR) frequencies and penalty analysis results [1] regarding dark chocolate attributes under different environments [2]. [1] Penalty analysis was associated with the overall liking scores (9-point hedonic scale). [2] D-PVR: dark chocolate positive VR; D-NVR: dark chocolate negative VR; D-B: dark chocolate sensory booth.

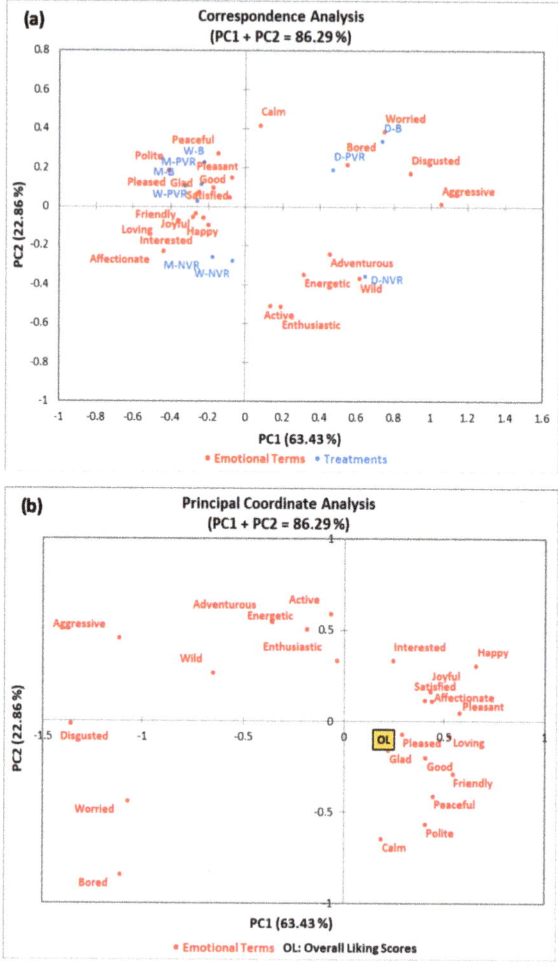

Figure 6. (**a**) Correspondence analysis (CA) of emotional terms for chocolate products tasted under different contextual settings [1]; (**b**) Principal coordinate analysis (PCoA) of emotional terms regarding the overall liking scores. [1] M-PVR: milk chocolate positive VR; M-NVR: milk chocolate negative VR; M-B: milk chocolate sensory booth; W-PVR: white chocolate positive VR; W-NVR: white chocolate negative VR; W-B: white chocolate sensory booth; D-PVR: dark chocolate positive VR; D-NVR: dark chocolate negative VR; D-B: dark chocolate sensory booth.

3.2.2. Principal Component and Cluster Analyses of the Chocolate Products under Different Environments

The principal component analysis (PCA) and agglomerative hierarchical clustering (HCA) results are shown in Figure 7. PCA biplot visualized the associations between liking scores of ten attributes and the three chocolate products (milk, white, and dark) while considering the contextual effect (Figure 7a). The principal component one (PC1) and principal component two (PC2) were 91.72% and 6.64%, respectively, explaining totally 98.36% of data variability. Liking vectors of most attributes were well linked with the horizontal axis, which was PC1 (squared cosines varied from 0.91 to 0.99). The liking vector of cocoa flavor was aligned with the vertical axis, which was PC2 (squared cosine was 0.53). Liking vectors of most attributes, except for cocoa flavor, were close to each other in Figure 7a, indicating their positive association. In addition, the liking vector of cocoa flavor was not associated with the liking vectors of hardness and texture as they were almost orthogonal. In terms of chocolate products, milk chocolate was highly associated with the liking of cocoa flavor under PVR and NVR, and milk chocolate under B was associated with the overall liking as well as the liking of bitterness, sweetness, smoothness, dairy flavor, aftertaste and taste/flavor. In addition, white chocolate was relatively associated with the liking of hardness and texture under PVR and NVR. However, dark chocolate was negatively correlated with the liking of all evaluated attributes regardless of the contextual effects.

Figure 7b shows the dendrogram based on AHC for the nine chocolate–environment combinations (3 × 3 factorial design). Three main cluster groups were formed, which were (1) dark chocolate under all environments, (2) milk chocolate under all environments, and (3) white chocolate under all environments.

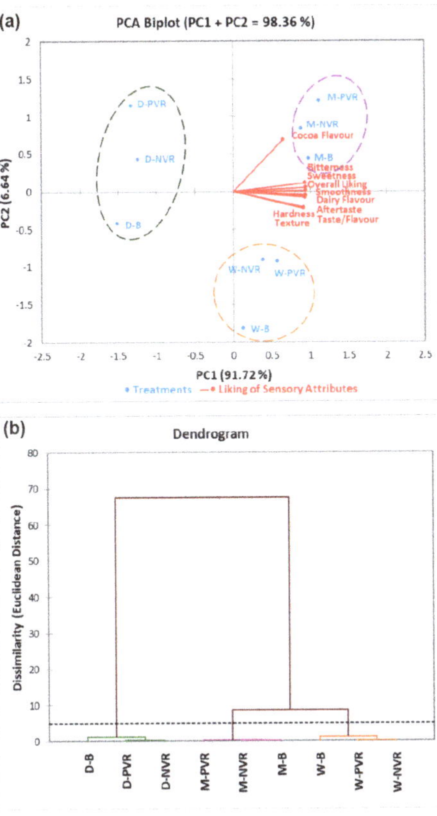

Figure 7. (**a**) Principal component analysis (PCA) biplot regarding liking scores [1] of chocolate attributes in different environments [2]; (**b**) Dendrogram of agglomerative hierarchical clustering (AHC) grouping chocolate products under different environments [2]. [1] Liking scores were based on the 9-point hedonic scale (1 = dislike extremely, 5 = neither like nor dislike, 9 = like extremely) [19]. [2] M-PVR: milk chocolate positive VR; M-NVR: milk chocolate negative VR; M-B: milk chocolate sensory booth; W-PVR: white chocolate positive VR; W-NVR: white chocolate negative VR; W-B: white chocolate sensory booth; D-PVR: dark chocolate positive VR; D-NVR: dark chocolate negative VR; D-B: dark chocolate sensory booth.

3.3. The Effect of Environments on the Purchase Intent of Chocolate Products

The frequencies of purchase intent for three chocolate products (milk, white, and dark) under the three environments (PVR, NVR, and B) are shown in Table 1. In general, milk chocolate had the highest purchase intent (64.2–73.1%) followed by white chocolate (56.7–62.7%), and dark chocolate had the lowest purchase intent (31.3–43.3%). Milk chocolate and white chocolate were not significantly different in their purchase intent under the three environments ($p > 0.05$). However, milk chocolate had significantly higher purchase intent than dark chocolate under the PVR, NVR, and B, respectively ($p < 0.05$). Based on the results, participants tended to be more willing to purchase three chocolate products under PVR, whereas there were no significant differences among PVR, NVR, and B regarding each chocolate type ($p > 0.05$).

Table 1. Purchase intent frequencies of chocolate products under different environments.

Treatments [1]		Willingness to Purchase (%) [2]
Chocolate	Environment	
Milk	PVR	73.1 [a]
	NVR	64.2 [ab]
	B	70.1 [a]
White	PVR	62.7 [ab]
	NVR	61.2 [ab]
	B	56.7 [abc]
Dark	PVR	43.3 [bc]
	NVR	31.3 [c]
	B	34.3 [c]

[1] three types of chocolate products (milk, white and dark) were tested under three contextual settings (PVR: positive VR, NVR: negative VR, and B: sensory booth). [2] Cochran's Q test was used together with Marascuilo procedure for multiple pairwise comparisons ($n = 67$); [a–c] Frequencies with different superscripts within the same column indicate significant differences (Cochran's Q test and Marascuilo procedure, $p < 0.05$).

4. Discussion

4.1. The Effect of Environments on Sensory Acceptability of Chocolate Products

4.1.1. Hedonic Ratings

The effect of context was found to not significantly affect the tasting experience of the chocolate types. The increased social element (via context) was found here not contributing enough to change the eating experience. Many reasons could be responsible for this finding, such as:

(a)   Large product effect with respect to empathy, [29], emotions [29], ecstasy, involvement or indulgent;
(b)   Irrelevant consumption context [30];
(c)   Familiarity with the product, thus context plays a smaller role [31,32];
(d)   Strong preference effect compared to context effect with respect to chocolate type, which may consequently help explain the stable preference pattern for chocolates, namely milk and white chocolate type;
(e)   Meal context, not testing context, is the more effective influence [31];
(f)   Attention bias could also make consumers focused only on chocolate testing and not to be affected by context environments, as consumers were aware of testing chocolate in three settings.

Pound and Duizer (2000) found similar no context effect results with chocolate type tested in four testing situations, namely, central location, in-home, teaching laboratory, and sensory laboratory. Similarly, context has previously been found to have no effect on the eating experience of cheese [30]. Thus, the effect of context depends on many of the abovementioned factors, which can contribute to the variability of the results.

In the present study, the type of chocolate was found to have significant effects on the hedonic ratings towards certain sensory attributes. Milk chocolate was the most liked product, followed by white chocolate, whereas the dark chocolate was the least liked among all the products (Figure 2). The three types of chocolate products used in this research varied in cocoa content, which were 28% (white), 33% (milk), and 72% (dark). Therefore, their sensory attributes can be largely affected by their ingredients, such as sweetness, bitterness, and cocoa flavor. Glicerina et al. [33] reported that these three chocolate types also have different textual properties. White chocolate has less aggregate structure and the lowest viscosity, whereas dark chocolate has the highest aggregate structure and fewer void spaces between particles. The microstructure and rheological properties of milk chocolate are in-between, which could be the reason for the highest liking scores. It has been reported that consumers prefer less hard and light chocolate products, which could be a support for the findings of the present research as well [34]. Apart from that, consumer preferences and relevant hedonic ratings may also be affected by demographic factors, such as gender [35]. Although the effect of the environment on the hedonic responses was marginal in this study, other reactions of consumers could be affected by the context in which consumers taste the product. As reported by Stelick and Dando [10], the environment where food products are consumed could affect the enjoyment, feelings, and purchase intent of alcoholic drinks [17].

4.1.2. JAR Results

JAR results can describe both the acceptability and intensity towards sensory attributes of products. In general, milk chocolate was the most acceptable product among the three chocolate types since its JAR selections for all attributes were comparatively high (Figure 3). However, the overall liking of milk chocolate was found to be affected by the contextual settings based on the penalty analysis results. The effect of an attribute on overall liking of the product is considered significant when the proportion of participants' responses to "not JAR" is greater than the commonly used threshold of 20% [36,37]. In the present study, both PVR and NVR led to the higher selection of "too little," and lower selections of JAR, and "too much" of the cocoa flavor for the milk chocolate (Figure 3). This might be because both PVR and NVR provided a better engagement than the sensory booth did, in which participants might focus more on the virtual experience than the chocolate itself. Bangcuyo et al. [15] also reported that consumers were more engaged in a coffee evaluation session that took place in an immersive virtual coffeehouse rather than in the traditional sensory booth. Therefore, the finding of this research might indicate that sensory evaluation conducted under immersive VR environments could have better engagement and ecological validity than traditional sensory booths.

The white and dark chocolates had lower JAR selections regarding all attributes compared to milk chocolate (Figures 4 and 5). About 81–87% of participants found that the overall texture of white chocolate under the three environments was just about right (Figure 4). Similar to the penalty analysis results, the other four attributes of white chocolate, including sweetness, bitterness, cocoa flavor, and dairy flavor, were penalized for being either "too much" or "too little" regardless of the environments. For the dark chocolate under three environments, all five attributes tended to have great negative effects on its overall liking (Figure 5). On the other hand, the sweetness and dairy flavor of white chocolate could be reduced and its cocoa flavor and bitterness could be increased for increasing consumers' liking, while on the contrary for dark chocolate. Overall, contextual settings did not affect the penalty analysis results for the white and dark chocolates. As previously mentioned, white chocolate has 28% cocoa content and less aggregate structure, whereas dark chocolate has 72% cocoa content and the highest aggregate structure [33]. In other words, both chocolate types have extreme cocoa content and textual properties, which could have greater effects than contextual settings (Figure 2). Milk chocolate has relatively moderate cocoa content (33%) and textual properties, which could minimize the effect of chocolate itself and enlarge the effect of contextual settings [33]. This is probably why the penalty analysis results for milk chocolate were different under the three environments.

4.2. Multivariate Analysis of Chocolate Products under Different Environments

4.2.1. Emotional Responses

In the present study, the most frequently selected emotional terms changed depending on both chocolate types as well as contextual settings (Figure 6). The descriptors selected for both milk chocolate and white chocolate were generally similar under different environments, which were both neutral and positive, such as "polite" and "affectionate." However, dark chocolate was associated with distinct emotions in different environments. Dark chocolate under NVR tended to have ardent emotional terms such as "adventurous" and "energetic." On the contrary, dark chocolate under both PVR and B were associated with negative terms, including "bored," "worried," "disgusted," and "aggressive," which contributed to low overall liking scores of chocolate products (<5.0). Overall, each chocolate product tasted under PVR elicited generally similar emotions as the control setting, namely sensory booths. However, NVR tended to evoke more passionate emotions from participants, especially towards dark chocolate.

Both VR environments used in this study showed their impacts on consumers' emotional responses towards chocolate products, especially the NVR. Xu et al. [38] reported that the environments where food products are consumed could heavily affect consumers' emotions. They found significant changes in emotions evoked from subjects when ice cream was consumed under laboratory, café, university study area, and bus stop settings. Apart from the visual effect, the auditory effect involved in this study should also be considered, as the VR environments were based on videos [39]. The liking of the two VR environments was highly subjective as dark chocolate was associated with negative emotions and low overall liking scores under B, which was the control setting, whereas the "virtual live concert" setting positively affected participants' emotional responses and overall acceptance of dark chocolate. According to the results, consumers might feel more appropriate to consume a dark chocolate rather than milk or white chocolates in a "live concert" environment.

4.2.2. Principal Component and Cluster Analyses of the Chocolate Products under Different Environments

Three clusters were formed primarily based on the type of chocolate products, whereas contextual settings were found not significant enough to affect the clustering in the present study (Figure 7). In general, milk chocolate was the most liked chocolate, and dark chocolate was the least liked chocolate based on the PCA results, which is similar to the finding in Figure 2. Both PVR and NVR were found to have positive effects on the cocoa flavor liking of milk chocolate and the textual liking of white chocolate. Some previous studies reported that both enjoyment and hedonic responses of food products tended to be higher under immersive VR environments [2,17]. However, those environments were pleasant and suitable for the consumption of relevant food products, such as tasting sparkling wine in a winery as well as vegetables in a holiday farm. Although the two VR environments used in this research belong to different categories, their positive effects towards certain chocolate attributes were similar. Hathaway and Simons [3] reported that the hedonic ratings of cookies obtained under relevant, immersive VR environments were more reliable and discriminating. Therefore, it would be necessary to select a matched tasting environment for each chocolate product in order to obtain reliable hedonic data. These sensory results could be further applied in testing newly developed chocolate products before launching.

4.3. The Effect of Environments on the Purchase Intent of Chocolate Products

In the present study, the purchase intent was highly associated with participants' hedonic responses as well as the PCA results, in which milk chocolate was the most liked chocolate type (Figures 2 and 7). As discussed above, milk chocolate had moderate cocoa content and textual properties, which could be the primary reason for the consumers' preference [33]. Although participants were most willing to pay for three chocolate products under PVR, there were no significant differences among

PVR, NVR, and B within each chocolate type ($p > 0.05$). Apart from the chocolate category, there are many factors that could affect consumers' willingness to purchase, such as the nutritional value and health claims of products [40]. Gunaratne et al. [26] also reported that the packaging could have effects on consumers' acceptance, purchase intent, and emotions regarding chocolate products. On the other hand, the consuming context may also contribute to those impacts. As reported by O'Brien and Toms [41], aesthetically pleasant environments could positively influence consumers' engagement. The highest frequencies of purchase intent achieved under PVR for each chocolate product might be correlated with improved engagement, which might have been perceived by the participants.

5. Conclusions

This preliminary study firstly explored the effect of 360-degree immersive videos (based on VR headsets) on the sensory perception of chocolate products. Changing the context environments did not significant affect the hedonic ratings of chocolates. As compared to the traditional sensory booth, participants tended to have better engagement when they tasted chocolate products under both VR environments. Therefore, the data could be more ecologically valid as well as relevant to actual consumers' experience. In addition, the combination of dark chocolate and "virtual live concert" significantly affected consumers' hedonic responses and emotions into a positive and passionate direction. However, it is necessary to further match each chocolate type to a suitable consuming environment achieved by VR headsets, for more reliable and ecologically valid sensory responses.

Supplementary Materials: The following is available online at http://www.mdpi.com/2304-8158/9/4/515/s1, Table S1: Frequencies and significance of emotional terms based on Cochran's Q test.

Author Contributions: Conceptualization, D.D.T. and Y.K.; Methodology, D.D.T. and Y.K.; Formal analysis, Y.K.; investigation, Y.K., M.K., M.T., L.L., and D.X.; data curation, Y.K.; writing—original draft preparation, Y.K.; Writing—review and editing, Y.K., D.D.T., C.S., and R.H.; Supervision, D.D.T., C.S., and R.H.; Project administration, D.D.T.; Funding acquisition, D.D.T. All authors have read and agreed to the published version of the manuscript.

Funding: This research was funded by Department of Wine, Food, and Molecular Biosciences, Faculty of Agriculture and Life Sciences, Lincoln University, New Zealand.

Acknowledgments: The authors would like to acknowledge all the subjects who participated in the sensory sessions.

Conflicts of Interest: The authors declare no conflict of interest. The funders had no role in the design of the study; in the collection, analyses, or interpretation of data; in the writing of the manuscript, or in the decision to publish the results.

References

1. Thurman, R.A. Instructional simulation from a cognitive psychology viewpoint. *Educ. Technol. Res. Dev.* **1993**, *41*, 75–89. [CrossRef]
2. Sinesio, F.; Saba, A.; Peparaio, M.; Civitelli, E.S.; Paoletti, F.; Moneta, E. Capturing consumer perception of vegetable freshness in a simulated real-life taste situation. *Food Res. Int.* **2018**, *105*, 764–771. [CrossRef] [PubMed]
3. Hathaway, D.; Simons, C.T. The impact of multiple immersion levels on data quality and panelist engagement for the evaluation of cookies under a preparation-based scenario. *Food Qual. Prefer.* **2017**, *57*, 114–125. [CrossRef]
4. Joiner, I.A. Virtual reality and augmented reality: What is your reality? In *Emerging Library Technologies*; Joiner, I.A., Ed.; Chandos Publishing: Oxford, UK, 2018; pp. 111–128.
5. Sherman, W.R.; Craig, A.B. *Understanding Virtual Reality: Interface, Application, and Design*; Morgan Kaufmann: Burlington, MA, USA, 2018.
6. Jia, J.; Chen, W. The ethical dilemmas of virtual reality application in entertainment. In Proceedings of the 2017 IEEE International Conference on Computational Science and Engineering and IEEE/IFIP International Conference on Embedded and Ubiquitous Computing, CSE and EUC, Guangzhou, China, 21–24 July 2017; pp. 696–699.

7. Smutny, P.; Babiuch, M.; Foltynek, P. A review of the virtual reality applications in education and training. In Proceedings of the 2019 20th International Carpathian Control Conference, ICCC, Krakow-Wieliczka, Poland, 26–29 May 2019.

8. Cikajlo, I.; Potisk, K.P. Advantages of using 3D virtual reality based training in persons with Parkinson's disease: A parallel study. *J. Neuroeng. Rehabil.* **2019**, *16*, 119. [CrossRef]

9. Loureiro, S.M.C.; Guerreiro, J.; Ali, F. 20 years of research on virtual reality and augmented reality in tourism context: A text-mining approach. *Tour. Manag.* **2020**, *77*, 104028. [CrossRef]

10. Stelick, A.; Dando, R. Thinking outside the booth—The eating environment, context and ecological validity in sensory and consumer research. *Curr. Opin. Food Sci.* **2018**, *21*, 26–31. [CrossRef]

11. Köster, E.P. Diversity in the determinants of food choice: A psychological perspective. *Food Qual. Prefer.* **2009**, *20*, 70–82.

12. Crofton, E.C.; Botinestean, C.; Fenelon, M.; Gallagher, E. Potential applications for virtual and Augmented Reality technologies in sensory science. *Innov. Food Sci. Emerg. Technol.* **2019**, *56*, 102178. [CrossRef]

13. Ares, G. Special issue on "Virtual reality and food: Applications in sensory and consumer science". *Food Res. Int.* **2019**, *117*, 1. [CrossRef]

14. Sester, C.; Deroy, O.; Sutan, A.; Galia, F.; Desmarchelier, J.-F.; Valentin, D.; Dacremont, C. "Having a drink in a bar": An immersive approach to explore the effects of context on drink choice. *Food Qual. Prefer.* **2013**, *28*, 23–31. [CrossRef]

15. Bangcuyo, R.G.; Smith, K.J.; Zumach, J.L.; Pierce, A.M.; Guttman, G.A.; Simons, C.T. The use of immersive technologies to improve consumer testing: The role of ecological validity, context and engagement in evaluating coffee. *Food Qual. Prefer.* **2015**, *41*, 84–95. [CrossRef]

16. Holthuysen, N.T.; Vrijhof, M.N.; de Wijk, R.A.; Kremer, S. "Welcome on board": Overall liking and just-about-right ratings of airplane meals in three different consumption contexts—laboratory, re-created airplane, and actual airplane. *J. Sens. Stud.* **2017**, *32*, e12254. [CrossRef]

17. Picket, B.; Dando, R. Environmental immersion's influence on hedonics, perceived appropriateness, and willingness to pay in alcoholic beverages. *Foods* **2019**, *8*, 42. [CrossRef] [PubMed]

18. Cardello, A.V.; Meiselman, H.L.; Schutz, H.G.; Craig, C.; Given, Z.; Lesher, L.L.; Eicher, S. Measuring emotional responses to foods and food names using questionnaires. *Food Qual. Prefer.* **2012**, *24*, 243–250. [CrossRef]

19. Peryam, D.R.; Pilgrim, F.J. Hedonic scale method of measuring food preferences. *Food Technol.* **1957**, *11*, 9–14.

20. Li, B.; Hayes, J.E.; Ziegler, G.R. Just-about-right and ideal scaling provide similar insights into the influence of sensory attributes on liking. *Food Qual. Prefer.* **2014**, *37*, 71–78. [CrossRef]

21. Ng, M.; Chaya, C.; Hort, J. Beyond liking: Comparing the measurement of emotional response using EsSense Profile and consumer defined check-all-that-apply methodologies. *Food Qual. Prefer.* **2013**, *28*, 193–205. [CrossRef]

22. Torrico, D.D.; Fuentes, S.; Viejo, C.G.; Ashman, H.; Gunaratne, N.M.; Gunaratne, T.M.; Dunshea, F.R. Images and chocolate stimuli affect physiological and affective responses of consumers: A cross-cultural study. *Food Qual. Prefer.* **2018**, *65*, 60–71. [CrossRef]

23. Demšar, J. Statistical comparisons of classifiers over multiple data sets. *J. Mach. Learn. Res.* **2006**, *7*, 1–30.

24. Pohlert, T. The pairwise multiple comparison of mean ranks package (PMCMR). *R Package* **2014**, *27*, 9.

25. Kourentzes, N. Tsutils: Time Series Exploration, Modelling and Forecasting. R Package Version 0.9. 0. 2019. Available online: https://CRAN.R-project.org/package=tsutils (accessed on 15 January 2020).

26. Gunaratne, N.M.; Fuentes, S.; Gunaratne, T.M.; Torrico, D.D.; Francis, C.; Ashman, H.; Gonzalez Viejo, C.; Dunshea, F.R. Effects of packaging design on sensory liking and willingness to purchase: A study using novel chocolate packaging. *Heliyon* **2019**, *5*. [CrossRef] [PubMed]

27. Gonçalves, S.; Ferreira, R.; Pereira, I.; Oliveira, C.C.D.; Amaral, P.I.S.; Garbossa, C.A.P.; Fonseca, L. Behavioral and physiological responses of different genetic lines of free-range broiler raised on a semi-intensive system. *J. Anim. Behav. Biometeorol.* **2017**, *5*, 112–117.

28. Palazzo, A.; Bolini, H. Sweeteners in diet chocolate ice cream: Penalty analysis and acceptance evaluation. *J. Food Stud.* **2017**, *6*, 13.

29. Van Boven, L.; Loewenstein, G.; Dunning, D.; Nordgren, L.F. Changing places: A dual judgment model of empathy gaps in emotional perspective taking. In *Advances in Experimental Social Psychology*; Elsevier: Amsterdam, The Netherlands, 2013; Volume 48, pp. 117–171.

30. Hersleth, M.; Ueland, Ø.; Allain, H.; Næs, T. Consumer acceptance of cheese, influence of different testing conditions. *Food Qual. Prefer.* **2005**, *16*, 103–110. [CrossRef]

31. Meiselman, H.L. Recent developments in consumer research of food. In *An Integrated Approach to New Food Product Development*; Moskowitz Sam, H.R.S., Straus, T., Eds.; CRC Press: Boca Raton, FL, USA, 2009; Volume 1, pp. 345–365.

32. Pound, C.; Duizer, L.; McDowell, K. Improved consumer product development. Part one. *Br. Food J.* **2000**, *102*, 810–820. [CrossRef]

33. Glicerina, V.; Balestra, F.; Dalla Rosa, M.; Romani, S. Microstructural and rheological characteristics of dark, milk and white chocolate: A comparative study. *J. Food Eng.* **2016**, *169*, 165–171. [CrossRef]

34. Kulozik, U.; Tolkach, A.; Bulca, S.; Hinrichs, J. The role of processing and matrix design in development and control of microstructures in dairy food production -A survey. *Int. Dairy J.* **2003**, *13*, 621–630. [CrossRef]

35. Zellner, D.A.; Garriga-Trillo, A.; Rohm, E.; Centeno, S.; Parker, S. Food liking and craving: A cross-cultural approach. *Appetite* **1999**, *33*, 61–70. [CrossRef]

36. Cusielo, K.V.C.; da Silva, A.C.D.M.L.; Tavares-Filho, E.R.; Bolini, H.M.A. Sensory influence of sweetener addition on traditional and decaffeinated espresso. *J. Food Sci.* **2019**, *84*, 2628–2637. [CrossRef]

37. Laguna, L.; Varela, P.; Salvador, A.; Fiszman, S. A new sensory tool to analyse the oral trajectory of biscuits with different fat and fibre contents. *Food Res. Int.* **2013**, *51*, 544–553. [CrossRef]

38. Xu, Y.; Hamid, N.; Shepherd, D.; Kantono, K.; Spence, C. Changes in flavour, emotion, and electrophysiological measurements when consuming chocolate ice cream in different eating environments. *Food Qual. Prefer.* **2019**, *77*, 191–205. [CrossRef]

39. Kantono, K.; Hamid, N.; Shepherd, D.; Lin, Y.H.T.; Brard, C.; Grazioli, G.; Thomas Carr, B. The effect of music on gelato perception in different eating contexts. *Food Res. Int.* **2018**, *113*, 43–56. [CrossRef] [PubMed]

40. Kaur, A.; Scarborough, P.; Rayner, M. A systematic review, and meta-analyses, of the impact of health-related claims on dietary choices. *Int. J. Behav. Nutr. Phys. Act.* **2017**, *14*. [CrossRef] [PubMed]

41. O'Brien, H.L.; Toms, E.G. The development and evaluation of a survey to measure user engagement. *JASIS* **2010**, *61*, 50–69. [CrossRef]

 *foods*

Article

Effects of Context and Virtual Reality Environments on the Wine Tasting Experience, Acceptability, and Emotional Responses of Consumers

Damir D. Torrico [1,2,*], Yitao Han [1], Chetan Sharma [2], Sigfredo Fuentes [1], Claudia Gonzalez Viejo [1] and Frank R. Dunshea [1]

[1] Faculty of Veterinary and Agricultural Sciences, School of Agriculture and Food, The University of Melbourne, Parkville VIC 3010, Australia; yitaoh1@student.unimelb.edu.au (Y.H.); sigfredo.fuentes@unimelb.edu.au (S.F.); claudia.gonzalez@unimelb.edu.au (C.G.V.); fdunshea@unimelb.edu.au (F.R.D.)

[2] Department of Wine, Food and Molecular Biosciences, Faculty of Agriculture and Life Sciences, Lincoln University, Lincoln 7647, New Zealand; Chetan.Sharma@lincoln.ac.nz

* Correspondence: Damir.Torrico@lincoln.ac.nz; Tel.: +64-34230641

Received: 24 January 2020; Accepted: 10 February 2020; Published: 14 February 2020

Abstract: Wine tasting is a multidimensional experience that includes contextual information from tasting environments. Formal sensory tastings are limited by the use of booths that lack ecological validity and engagement. Virtual reality (VR) can overcome this limitation by simulating different environmental contexts. Perception, sensory acceptability, and emotional responses of a Cabernet Sauvignon wine under traditional sensory booths, contextual environments, and VR simulations were evaluated and compared. Participants ($N = 53$) performed evaluations under five conditions: (1) traditional booths, (2) bright-restaurant (real environment with bright lights), (3) dark-restaurant (real environment with dimly lit candles), (4) bright-VR (VR restaurant with bright lights), and (5) dark-VR (VR restaurant with dimly lit candles). Participants rated the acceptability of aroma, sweetness, acidity, astringency, mouthfeel, aftertaste, and overall liking (9-point hedonic scale), and intensities of sweetness, acidity, and astringency (15-point unstructured line-scale). Results showed that context (booths, real, or VR) affected the perception of the wine's floral aroma (dark-VR = 8.6 vs. booths = 7.5). Liking of the sensory attributes did not change under different environmental conditions. Emotional responses under bright-VR were associated with "free", "glad", and "enthusiastic"; however, under traditional booths, they were related to "polite" and "secure". "Nostalgic" and "daring" were associated with dark-VR. VR can be used to understand contextual effects on consumer perceptions.

Keywords: virtual reality; acceptability; Cabernet Sauvignon; wine; context; emotions

1. Introduction

The globalization of markets, an increase in per capita income (especially in developing countries), and eventually increased spending power has allowed more consumers to access wine products from around the world [1]. With the increasing consumers' wine demand and increased competition, the production of higher quality and more acceptable wines is one of the biggest challenges for the wine industry to remain relevant in the market [2]. Sensory analysis of wines is an essential and critical component of quality control [3,4]. The chemical composition measured by instrumental analysis can provide valuable and reliable information to assess the quality of wine products; however, the consumers' assessment is a critical decision-making tool to test the success of the product in the marketplace [1,2]. Sensory analysis, which relies on the sight, smell, taste, touch, and hearing senses to determine the intrinsic properties of wine products provides a multidimensional response that is closely related to the tasting experience of consumers. From the vineyards (grape quality) to the final wine products that are being consumed on various occasions, sensory analysis is an important tool in the manufacturing chain of wine production and commercialization [5].

Traditional wine assessment depends on the tasting abilities of a panel of experts who rate the quality of wines using an attribute-based grading system [2]. However, this method only uses a small number of scores that might not explain the acceptability of the product entirely. On the other hand, standard consumers' sensory evaluation relies on using isolated booths for removing external environmental factors, including odors and noises that can produce biases in responses [6,7]. However, these controlled environments lack ecological validity because they do not offer "real-world" conditions for replicating the authentic tasting experience of consumers [8]. Sensory laboratories can be considered unfamiliar environments for the consumption of foods and beverages, where consumers are separated in booths and isolated from external stimuli. In reality, consumption of foods and beverages is influenced by the environmental factors that stimulate the senses of consumers [9–11]. The contextual elements are complex and variable, including several sensory stimuli, which constitute the background information of the tasting experience. The different external factors create complex contextual cues that are essential for the formation of subsequent consumers' behaviors, expectations, and hedonic evaluations [12–15]. Besides the sensory characteristics of products, external factors surrounding consumers (environmental and social interaction) can also affect the acceptability and emotional responses [16].

The lack of active consumers' engagement in the sensory and hedonic evaluations can reduce the prediction rates of product choice, which leads to a higher probability of failure in the research and development of new products in the food and beverage industry [8]. The primary purpose of using a sensory laboratory is to collect responses generated by the senses of participants with the elimination of the influences produced by external factors [17]. Compared to traditional sensory laboratory setups, "real-world" environments have multiple external variables that are difficult to quantify and control. For instance, Dorado, Chaya, Tarrega, and Hort [16] stated that the contextual information could affect the emotional responses of consumers towards beer products, but this effect was not significant for liking. Sester et al. [18] found that different contexts, produced by the simulation of two bar environments using immersive technology (bars were decorated with cold and warm tone lights with different background pictures), had a significant effect on consumers' choices of drinks. Nijman et al. [19] found that consumers were able to discriminate lager and ale beers better using a real bar as a testing environment compared to that of traditional sensory booths. However, conducting sensory research in external locations is challenging because it can be generally time-consuming and expensive [20].

The creation of a simulated virtual environment within the controllable laboratory setup for the sensory evaluation of wine products could be a step closer to pursue ecological validity at a relatively lower cost. Virtual reality (VR) technology, which combines a series of interactive computer-generated images or videos that link users' minds and sensory systems, can be applied to simulate environments similar to the "real world" [21]. The virtual environment provided by the VR technology usually is generated by using VR headsets, which can offer visual and auditory stimuli with the advantage of being easy to operate [22]. VR technology can generate virtual scenes throughout dynamic environments that make participants feel engaged by using stereoscopic displays and sensing technology. Moreover, compared with the "real-world" environment, the conditions of VR environments are relatively controllable, and the environments are easy to replicate. There is a growing interest in the scientific community on studying the effects of VR on sensory and consumer sciences [23–25]. Regarding the use of real surroundings, Hannum et al. [26] evaluated the effect of three contextual environments (traditional sensory booths, an immersive wine bar, and an actual wine bar) on the acceptability and purchase intent of consumers toward wine samples. Although they found that the environment had a marginal effect on acceptability, individual consumers' behaviors changed depending on the environment that was used. There is still very little information on how the VR environments perform against "real-world" contexts. Therefore, this study evaluated the perception, sensory acceptability, and emotional responses of a Cabernet Sauvignon wine under different conditions, including traditional sensory booths, "real-word" contextual environments, and VR simulations with associated hardware.

2. Materials and Methods

2.1. Participants

The research protocol for this study was listed as a minimal risk with the ethics approval 1543704.2 obtained in February 2017 by the Human Ethics Advisory Group (HEAG) of the Faculty of Veterinary and Agricultural Science at The University of Melbourne, Australia. A total of $N = 53$ participants (12 males and 41 females) with ages of 35 ± 10 years old were recruited from a pool of students and staff belonging to The University of Melbourne to participate in this study. A power analysis was performed to verify that the number of participants (N) used in the present study was adequate (*Power* $= 0.93$). All participants were untrained and reportedly not allergic to any food product. Participants who consumed wine products at least once per month were pre-selected for the sensory sessions. All participants were asked to sign a consent form approved by the Human Ethics Advisory Group (The University of Melbourne) before tasting the wine. Sensory sessions were held at the sensory laboratory facilities of The University of Melbourne, Australia, which is composed of 20 sensory booths and focus group rooms. Five sensory sessions (traditional booths, bright-restaurant, dark-restaurant, bright-VR, and dark-VR) were conducted on five different days. The order of the sessions was randomized within each participant. The duration of one sensory session was approximately 20 to 30 min for each participant. At the end of all five sessions, participants were rewarded with a $10 gift card.

2.2. Stimulus

A descriptive panel ($N = 8$) from The University of Melbourne was used to select the Cabernet Sauvignon wine for this study. Six red wines were classified as high, medium, and low according to their average quality scores (100 points). A medium-quality (*score* $= 76.25$) Cabernet Sauvignon wine (Phoenix, Penley Estate, Coonawarra, SA, Australia) was used as the product stimulus for this experiment. Only one medium-quality red wine was selected because the focus of this experiment was to measure the effects of the context and environment on the taste perception of the wine. Bottles of 750 cc wines from the same batch were purchased from a local grocery store and kept at 16 °C. Five hours before the sensory sessions, the bottles of wines were transferred to the tasting room to reach room temperature (25 °C). Wine bottles were wrapped with aluminum foil to hide any packaging cues that can bias the responses of participants. A total of 15 mL of wine was poured in standard wine clear glasses (international standard wine tasting glasses, Luigi Bormioli International Organization for Standardization (ISO) wine tasting glasses with a rim diameter of 46 mm, height of 155 mm, and a total volume of 215 mL). For each environment, participants received two wine samples with different three-digit random codes from the same bottle (duplicates). This was done to avoid positional biases, logical errors, and pattern effects that can lead to guessing the nature of the samples. The purpose of this test was to measure the effect of the context and environment on consumers' responses having samples with the same taste.

2.3. Sensory Procedure

At the beginning of the tasting sessions, instructions were provided to each participant explaining the experimental procedures, including the proper operation and wearing of VR headsets, as well as information about how to fill out the paper ballots. After a brief explanation of the test procedures, participants who were willing to continue with the sensory test signed the consent form. The environments in this study included (1) traditional booths, (2) bright-restaurant (real environment illuminated with bright lights), (3) dark-restaurant (real environment illuminated with dimly lit candles), (4) bright-VR (VR restaurant illuminated with bright lights), and (5) dark-VR (VR restaurant illuminated with dimly lit candles) (Figure 1). In the VR sessions, participants were instructed to wear the VR headsets and taste the wine with the help of the instructor in the room. For the immersive (real) and traditional booth sessions, participants were seated in the tables and had the samples ready in front of them for tasting. Participants tasted two wine samples (using the same wine) in each session. The presentation of the samples was randomized, and a sequential monadic sample order was used within each participant. In the VR sessions, participants were instructed to take VR headsets off and start answering questions once they finished tasting. This process was repeated for each participant until the tasting of all samples was completed under each VR setting. In the immersive (real) and traditional booth sessions, participants were instructed to taste the wine samples from left to right and answer the questions in the paper ballots. In the paper ballots, participants were asked to rate the acceptability of the floral aroma, fruity aroma, sweetness, acidity, mouthfeel-body, astringency, aftertaste, and overall liking of the red wine sample using the nine-point hedonic scale (1 = disliked extremely, 5 = neither liked nor disliked, 9 = liked extremely). The intensities of floral aroma, sweetness, acidity, and astringency were evaluated using a 15 cm unstructured line scale. Sweetness, acidity, astringency, and body were also assessed using a just-about-right scale (JAR; for sweetness, acidity, and astringency: 1 = too little, 2 = just about right, 3 = too much; for body: 1 = light, 2 = medium, 3 = full). Purchase intent (question: "If this product is available on the market, will you buy it?") of the wine samples was evaluated using a binomial scale (1 = yes, 2 = no). No re-tasting of samples was allowed in the current experiment, and all the collected responses were memory-based.

To assess the emotional responses elicited by the wine tasting experience in the different environments, a check-all-that-apply (CATA) procedure was applied using a list of 33 emotional terms (adventurous, pleased, satisfied, pleasant, active, secure, affectionate, warm, calm, bored, energetic, disgusted, enthusiastic, worried, free, aggressive, friendly, daring, glad, eager, good, guilty, happy, polite, interested, steady, joyful, understanding, loving, wild, merry, nostalgic, and peaceful).

These emotion terms were pre-selected from a list containing 48 emotional terms obtained from previous studies [27,28] to cover two-dimensional affective spaces (valence and arousal), according to Bradley and Lang [29]. Participants used water and unsalted crackers to cleanse their palate in between wine samples.

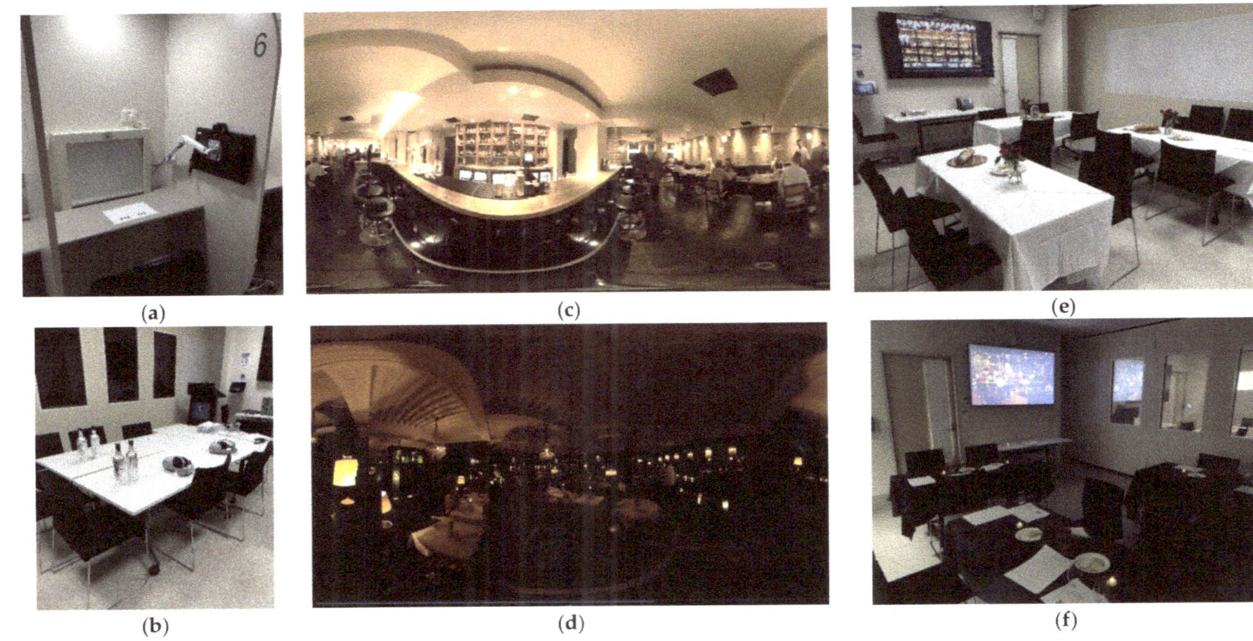

Figure 1. Experimental settings [1] for the sensory evaluation of the wine. [1] (a) Traditional sensory booths; (b) VR set-up; (c) bright restaurant VR environment; (d) dark restaurant VR environment; (e) bright real restaurant environment; and (f) dark real restaurant environment.

2.4. Test Environments

As shown in Figure 1, five test environments (traditional booths, bright-restaurant, dark-restaurant, bright-VR, and dark-VR) were used for the tasting experiences of the wine. Traditional booths (traditional sensory evaluation environment) consisted of isolated individual booths located in the sensory laboratory facilities at the University of Melbourne, Australia (Figure 1a). The dimensions of the sensory testing booths were 1.5 m (width) × 2.1 m (height) with a worktop used for placing samples and questionnaires. The sensory booths were illuminated with light emitting diode (LED) light (configured with white color; RGB = 255, 255, 255). The temperature of the sensory room was set at 25 °C. The VR environments used in this study (restaurants with bright and dark ambient) were both obtained from YouTube (Google LLC, San Bruno, CA, USA), and were selected from a pool of 10 VR environments (YouTube) in preliminary focus group discussions (N = 6) (bright-VR contextual environment (https://www.youtube.com/watch?v=y8iqpLN-YIE), and dark-VR contextual environment (https://www.youtube.com/watch?v=2zYWdAqmxBw)) The "real" restaurant environments were assembled to create two restaurant conditions for this study (bright and dark conditions). Before the testing session, wine samples coded with three-digit random numbers and questionnaires (paper ballots) were placed on the table for the participants.

Consumer tests in the "real" restaurant contexts were conducted in two separate rooms having bright and dark environments (Figure 1e,f, respectively). In the bright-restaurant environment, participants were seated in restaurant-type tables that were decorated with a vase with flowers and a wooden table with bread. In the bright-restaurant room, there was a 60 inch TV screen (Samsung, Seoul, Korea) that displayed a photo of a wine cabinet with a wide variety of wines. In the dark-restaurant environment, participants were seated in a dark room that was illuminated by artificial LED tea-lights (Kmart, Melbourne, VIC, Australia), and decorated with flowers and plates. In the dark-restaurant room, there was also a 60 inch TV screen that displayed a photo of a dark-restaurant room.

The consumer test sessions under the VR environments were conducted in a private and isolated focus-group room (Figure 1b). The VR environments were generated by an Oculus Go all-in-one VR headset with a controller (Facebook Technologies, LLC, Menlo Park, CA, USA), which provided the dynamic visual scenarios (Figure 1c,d). Throughout the VR testing process, a testing supervisor was always present to help the subjects wear the VR headsets. After the tasting of each wine sample, participants were asked to remove their VR headsets and answer the questions on the paper ballots. The bright-restaurant VR environment was a restaurant-type room illuminated with bright lights. The subjects in this environment tasted the wines in a simulated restaurant bar, facing a cabinet that has a wide variety of wines (Figure 1c). During the tasting experience, participants were able to see, hear, and feel the activities of other persons dining in the restaurant. The background noise for this environment included low-pitched conversations from other persons in the restaurant. On the other hand, the dark-restaurant VR was placed in a restaurant illuminated with dim lights with no background music. The restaurant had several tables with seated persons that were having low-pitched conversations. Table lamps with yellow lights were placed in several places in the restaurant. Participants in this VR environment were placed in one table at the corner of the restaurant (Figure 1d).

2.5. Statistical Analysis

A two-way ANOVA (environment × order) with a generalized linear model (GLM) and a post-hoc Tukey's Honest Significant Difference (HSD) test were used to asses significant differences among the evaluated environments (traditional booths, bright-restaurant, dark-restaurant, bright-VR, and dark-VR) for the hedonic ratings and intensity scores of the Cabernet Sauvignon wine. The order effect was included in the ANOVA model to test whether the position of the samples was a significant factor when tasting the wine. A penalty test on the JAR data was performed to determine the effects of sensory attributes on the hedonic liking. Mean drops for the "too much" and "too little" scores were calculated (differences between the liking mean for the JAR level minus the "too much" or "too little" levels). For the CATA frequency data, correspondence analysis and principal coordinate analysis were used to assess the differences among the evaluated environments relative to the selection of the emotion terms and overall liking levels. For the purchase intent, the Cochran's Q test and simultaneous confidence intervals testing were used for multiple comparisons. A principal component analysis (PCA) was applied to interpret relationships between the hedonic ratings and intensity scores of the wine in different environments. A product-attribute biplot was used for the illustration of the PCA. Data were analyzed at $\alpha = 0.05$ using the XLSTAT Statistical Software version 2017 (Addinsoft, New York, NY, USA). All data were reported as mean values with standard errors.

3. Results

3.1. Sensory Responses to the Wine Sample in Different Environments

Results of the analysis of variance (ANOVA) for the different sensory parameters are shown in Table 1 (acceptability and intensity). For the acceptability responses (floral aroma, fruity aroma, sweetness, acidity, mouthfeel astringency, body, aftertaste, and overall liking), none of the main effects (environment nor order) were significant ($p \geq 0.05$) in the ANOVA models. The interaction (environment*order) effect was only significant ($p < 0.05$) for the fruity aroma attribute on the model of acceptability. For the intensity parameters (floral aroma, sweetness, acidity, and astringency), the type of environment effect was only significant ($p < 0.05$) for the floral aroma and astringency parameters. The interaction effect (environment*order) was only significant ($p < 0.05$) for the acidity and astringency intensity parameters. The order main effect was not a significant factor ($p \geq 0.05$) for neither the acceptability nor the intensity parameters; therefore, the means of the two served samples could be pooled for the *post-hoc* means comparison analysis.

Table 1. ANOVA [1] table for the acceptability and intensity parameters of the wine by environment samples.

	ANOVA Environments							
	Acceptability							
Effects [1]	**Floral Aroma**		**Fruity Aroma**		**Sweetness**		**Acidity**	
	F Value [2]	Pr > *F* [2]	*F* Value [2]	Pr > *F* [2]	*F* Value [2]	Pr > *F* [2]	*F* Value [2]	Pr > *F* [2]
Environment	1.15	0.33	1.25	0.29	1.09	0.36	0.92	0.45
Order	0.46	0.50	1.61	0.20	0.00	0.96	0.05	0.83
Environment*Order [3]	1.44	0.22	2.63	*0.03*	1.66	0.16	0.81	0.52
	Acceptability							
Effects [1]	**Mouthfeel Astringency**		**Mouthfeel Body**		**Aftertaste**		**Overall Liking**	
	F Value [2]	Pr > *F* [2]	*F* Value [2]	Pr > *F* [2]	*F* Value [2]	Pr > *F* [2]	*F* Value [2]	Pr > *F* [2]
Environment	0.76	0.55	0.28	0.89	0.16	0.96	0.43	0.79
Order	0.33	0.57	0.68	0.41	2.05	0.15	0.47	0.49
Environment*Order [3]	0.99	0.41	0.87	0.48	0.85	0.50	1.53	0.19
	Intensity							
Effects [1]	**Floral Aroma**		**Sweetness**		**Acidity**		**Astringency**	
	F Value [2]	Pr > *F* [2]	*F* Value [2]	Pr > *F* [2]	*F* Value [2]	Pr > *F* [2]	*F* Value [2]	Pr > *F* [2]
Environment	4.04	*<0.01*	0.73	0.57	1.05	0.38	2.42	*0.05*
Order	0.02	0.89	2.41	0.12	0.47	0.49	0.03	0.86
Environment*Order [3]	1.42	0.23	1.11	0.35	2.93	*0.02*	5.08	*<0.01*

[1] ANOVA = analysis of variance (five types of environments (traditional booths, bright-restaurant, dark-restaurant, bright-VR, and dark-VR) and two positional orders. Liking scores were based on a nine-point hedonic scale (1 = dislike extremely, 9 = like extremely). Intensity scores were based on a 15-point Likert scale (1 = absent, 15 = very strong). [2] *F* value, mean square/mean square error. Effects were considered significant when the probability Pr > *F* was less than 0.05 (bolded and italicized probabilities). [3] The environment effect was crossed with the replicate effect in a two-way factorial design (type of environment by order) using participants as blocks.

Table 2 shows the mean acceptability scores of the red wine sample in each environmental condition (traditional booths, bright-restaurant, dark-restaurant, bright-VR, and dark-VR). For the aroma acceptability attributes, the floral aroma scores for all the environments were not significantly different ($p \geq 0.05$) among them, ranging from 5.65 (dark-VR) to 5.94 (dark-restaurant). On the other hand, the acceptability score of the fruity aroma of the wine sample in the dark-restaurant environment was significantly higher ($p < 0.05$) compared to that of the wine sample in the traditional booths (5.99 vs. 5.65, respectively). No significant differences ($p \geq 0.05$) were found among the environments in the taste acceptability parameters (sweetness, acidity, astringency, body, and aftertaste) and overall liking of the wine sample. The mean intensity scores of the wine sample in each environment for the floral aroma, sweetness, acidity, and astringency are also shown in Table 2. The real dark-restaurant environment had a significantly ($p < 0.05$) higher floral aroma intensity score compared to those values of the traditional booths and the VR environments (bright and dark; 8.61 vs. 7.45–7.96). On the other hand, the real bright-restaurant had a significantly ($p < 0.05$) higher floral aroma intensity score compared to the value of the traditional booths (8.14 vs. 7.45, respectively), but was not significantly ($p \geq 0.05$) different compared to the VR environments (bright and dark). Opposite results were observed for the astringency intensity attribute, in which the real dark-restaurant environment had a significantly ($p < 0.05$) lower score compared to that of the traditional booths and the dark-VR environment (7.57 vs. 8.34–8.39, respectively; Table 2). For the sweetness and acidity intensity attributes, no significant differences ($p \geq 0.05$) were found among the environments.

Table 2. Acceptability and intensity mean values of the wine sample in each environment [1].

Environments [1]	Acceptability [2]			
	Floral Aroma	Fruity Aroma	Sweetness	Acidity
Traditional booths	5.66 ± 1.59 [a]	5.65 ± 1.58 [b]	5.26 ± 1.68 [a]	5.06 ± 1.71 [a]
Bright-restaurant	5.90 ± 1.36 [a]	5.91 ± 1.38 [a, b]	5.19 ± 1.77 [a]	5.21 ± 1.78 [a]
Dark-restaurant	5.94 ± 1.49 [a]	5.99 ± 1.35 [a]	5.41 ± 1.66 [a]	5.23 ± 1.69 [a]
Bright-VR	5.81 ± 1.32 [a]	5.79 ± 1.31 [a, b]	5.47 ± 1.72 [a]	5.10 ± 1.65 [a]
Dark-VR	5.65 ± 1.49 [a]	5.90 ± 1.44 [a, b]	5.59 ± 1.77 [a]	5.44 ± 1.66 [a]
Environments [1]	Acceptability [2]			
	Astringency	Body	Aftertaste	Overall Liking
Traditional booths	5.15 ± 1.81 [a]	5.67 ± 1.47 [a]	5.59 ± 1.49 [a]	5.57 ± 1.60 [a]
Bright-restaurant	5.48 ± 1.54 [a]	5.73 ± 1.36 [a]	5.49 ± 1.37 [a]	5.58 ± 1.46 [a]
Dark-restaurant	5.41 ± 1.85 [a]	5.58 ± 1.57 [a]	5.58 ± 1.55 [a]	5.60 ± 1.66 [a]
Bright-VR	5.28 ± 1.43 [a]	5.62 ± 1.31 [a]	5.50 ± 1.50 [a]	5.54 ± 1.52 [a]
Dark-VR	5.44 ± 1.79 [a]	5.75 ± 1.54 [a]	5.60 ± 1.59 [a]	5.77 ± 1.56 [a]
Environments [1]	Intensity [2]			
	Floral Aroma	Sweetness	Acidity	Astringency
Traditional booths	7.45 ± 3.04 [c]	6.62 ± 2.88 [a]	8.16 ± 2.67 [a]	8.39 ± 2.14 [a]
Bright-restaurant	8.14 ± 3.13 [a, b]	6.67 ± 2.76 [a]	8.25 ± 2.84 [a]	8.23 ± 2.55 [a]
Dark-restaurant	8.61 ± 2.93 [a]	6.97 ± 2.71 [a]	8.09 ± 2.44 [a]	7.57 ± 2.62 [b]
Bright-VR	7.96 ± 2.65 [b, c]	7.05 ± 3.00 [a]	7.65 ± 2.57 [a]	7.78 ± 2.79 [a, b]
Dark-VR	7.84 ± 2.84 [b, c]	7.01 ± 2.93 [a]	7.94 ± 2.80 [a]	8.34 ± 2.59 [a]

[1] Five environments were tested (traditional booths, bright-restaurant, dark-restaurant, bright-VR, and dark-VR). Means and standard deviations of 53 participants. [2] Liking scores were based on a nine-point hedonic scale (1 = dislike extremely, 9 = like extremely). Intensity scores were based on a 15-point Likert scale (1 = absent, 15 = very strong). [a–c] Means with different superscripts in each column within each attribute indicate significant differences ($p < 0.05$) by the Tukey studentized range Honest Significant Difference (HSD) test.

The frequency distribution (%) of participants' responses for the intensities of sweetness, acidity, astringency, and body of the wine sample in each environment using the just-about-right (JAR) scale is shown in Figure 2. In general, the wine sample was considered to be "just-about-right" (49%–59%) and "too little" (37%–50%) in the sweetness for all tested environments in this study. Moreover, participants considered the wine sample to be "just-about-right" (52%–66%) and "too much" (28%–38%) in acidity and astringency for all environments. For the body of the wine sample, participants rated this attribute as "just-about-right" (61%–71%) for all the environments (Figure 2). The penalty analysis using the JAR data is shown in Table 3. In general, the wine sample for all the environments was considered to be "too little" in sweetness (*mean drop* = 1.25–1.99; $p < 0.05$) except for the dark-VR environment, in which the mean drop was not significant in the overall liking (0.40; $p \geq 0.05$).

Table 3. Penalty analysis results for the sweetness, acidity, astringency, and body of the wine sample in different environments [1].

Environment	Sweetness		Acidity		Astringency		Body	
	Too Little	Too Much	Too Little	Too Much	Too Little	Too Much	Too Little	Too Much
Booths	*1.54*	−0.25	1.66	*1.44*	0.14	*1.52*	0.72	0.17
Bright-real	*1.25*	0.21	0.43	*1.30*	0.01	*1.34*	−0.20	−0.67
Dark-real	*1.99*	0.06	1.58	*1.94*	0.31	*1.94*	−0.08	−0.02
Bright-VR	*1.51*	0.60	0.89	*1.17*	1.47	*1.76*	0.66	0.01
Dark-VR	0.40	−1.64	−0.47	*1.24*	−1.06	*1.02*	−1.35	1.00

[1] Booths = traditional sensory booths, Bright-real = bright restaurant real environment, Dark-real = dark restaurant real environment, Bright-VR = bright restaurant VR environment, and Dark-VR = dark restaurant VR environment. Values represent the mean drops using the nine-point hedonic scale. Mean drops were considered significant when the probability Pr > F was less than 0.05 (bolded and italicized values).

Participants penalized the wine sample in all the environments for being "too much" in acidity (*mean drop* = 1.17–1.94) and astringency (1.02–1.94), but they did not penalize the body attribute ($p \geq 0.05$; Table 3). Moreover, the purchase intent values of the wine samples in all the environments were not significantly ($p \geq 0.05$) different (42%–45%; data not shown).

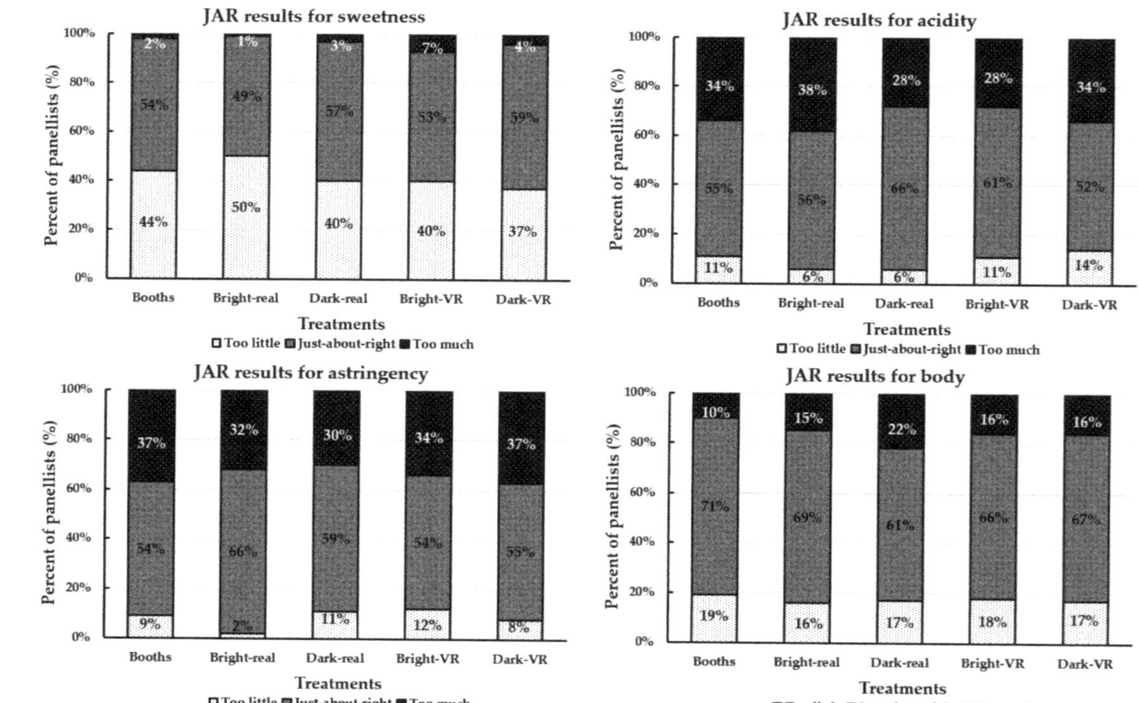

Figure 2. Selection frequencies (%) of just-about-right (JAR) results for the sweetness, acidity, astringency, and body of the wine sample in different environments [1]. [1] Booths = traditional sensory booths, Bright-real = bright restaurant real environment, Dark-real = dark restaurant real environment, Bright-VR = bright restaurant virtual reality (VR) environment, and Dark-VR = dark restaurant VR environment.

3.2. Emotions and Multivariate Analysis of the Wine Sample in Different Environments

Figure 3a shows the corresponding analysis of the emotional terms of the CATA question related to the wine sample in each environment. The principal component one (PC1) and principal component two (PC2) accounted for 26.41% and 36.65%, respectively, which explained 63.03% of the total data variability. The wine sample was only associated with the emotions "polite" and "calm" under the traditional booths environment. Under the real bright-restaurant environment, participants' emotions toward the wine sample were associated with "interested", "secure", "friendly", and "loving".

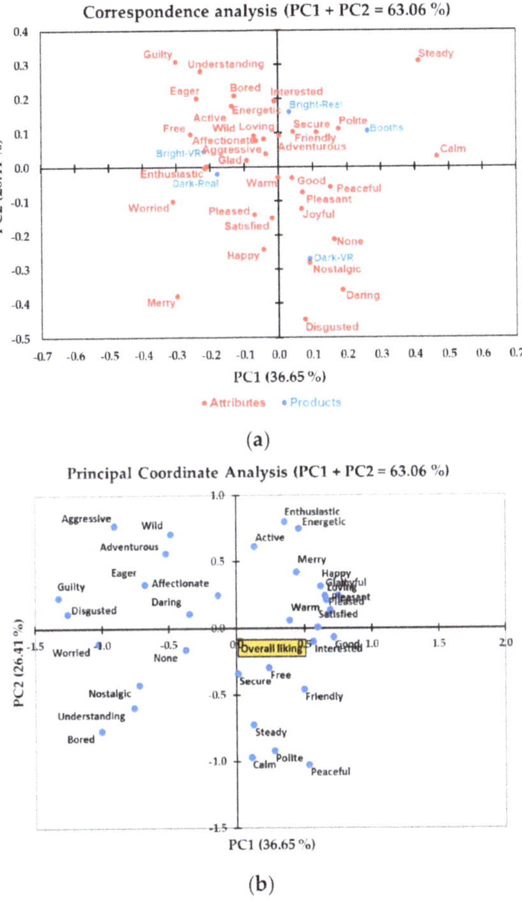

(a)

(b)

Figure 3. (a) Correspondence analysis of the emotion terms for the wine sample in each environment [1] and (b) principal coordinate analysis of the emotion terms with the overall liking score. [1] Booths = traditional sensory booths, Bright-real = bright restaurant real environment, Dark-real = dark restaurant real environment, Bright-VR = bright restaurant VR environment, and Dark-VR = dark restaurant VR environment.

On the other hand, participants only elicited emotions such as "free" and "enthusiastic" towards the wine sample under the bright-VR environment. The emotions "glad" and "enthusiastic" were associated with the wine under the real dark-restaurant environment. Conversely, "nostalgic", "daring", and "disgusted" were associated with the wine under the dark-VR environment. The principal coordinate analysis of the emotion terms concerning the overall liking of the wine sample in different environments is shown in Figure 3b. In general, overall liking was positively associated with the emotion terms "secure", "free", "interested", "good", and "friendly." Moreover, the overall liking of the samples was negatively associated with "daring", "affectionate", "eager", "adventurous", and "wild."

The principal component analysis (PCA) biplot shows the acceptability and intensity parameter vectors associated with the five environmental conditions (traditional booths, bright-restaurant, dark-restaurant, bright-VR, and dark-VR) in which the wine sample was tasted (Figure 4). Considering all acceptability and intensity sensory parameters, the PC1 and PC2 accounted for 36.93% and 31.77% of the biplot, respectively, explaining 68.7% of the total data variability. The fruit aroma liking and floral aroma intensity (*factor loading* = 0.89–0.91; data not shown) vectors contributed largely to the discrimination of the environments in the PC1.

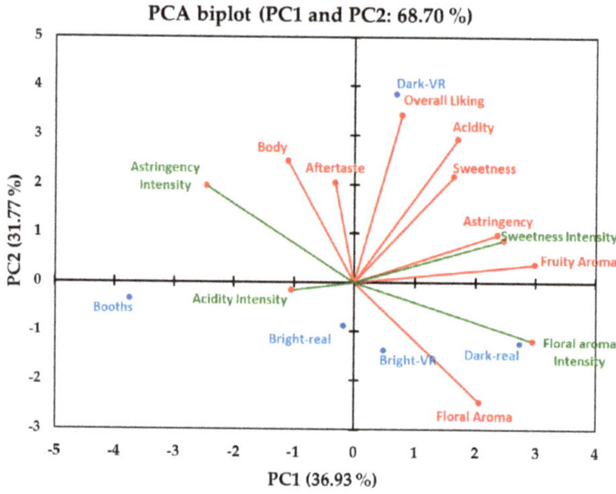

Figure 4. Principal component analysis (PCA) biplot visualizing treatments [1] (wine sample in each environment), acceptability (liking; vectors in red color), and intensity attributes (vectors in green color). [1] Booths = traditional sensory booths, Bright-real = bright restaurant real environment, Dark-real = dark restaurant real environment, Bright-VR = bright restaurant VR environment, and Dark-VR = dark restaurant VR environment.

On the other hand, the overall liking vector (*factor loadings* = 0.96) contributed largely to the discrimination of the samples in the PC2. According to the PCA, the acidity, sweetness, and aftertaste liking scores were positively associated with overall liking. On the other hand, the liking of the floral aroma was positively associated with the intensity of the floral aroma, but it was negatively associated with the intensity of astringency. The sweetness intensity was positively associated with the liking of the fruity aroma and astringency, but it was negatively associated with the intensity of the acidity. Moreover, the liking of the body was positively associated with the liking of the aftertaste, but it was negatively associated with the liking of the floral aroma. The real dark-restaurant and bright-VR environments were related to higher floral aroma intensity and liking. The dark-VR environment was related to a higher overall liking score, and the traditional booths environment was related to a higher intensity of acidity (Figure 4).

4. Discussion

4.1. Sensory Responses to the Wine Sample in Different Environments

This study showed that the type of environment (traditional booths, bright-restaurant, dark-restaurant, bright-VR, and dark-VR) had a marginal effect on the sensory acceptability (floral aroma, sweetness, acidity, astringency, body, aftertaste, and overall liking) of the Cabernet Sauvignon wine sample (Tables 1 and 2). Only the real dark-restaurant environment had a higher acceptability score for the fruity aroma attribute compared to that of the traditional booths (5.99 vs. 5.65, respectively; Table 2). On the other hand, the type of environment significantly affected the intensity perception of the floral aroma and astringency of the wine sample. The wine tasted in both real environments (bright and dark) had a significantly higher floral aroma compared to that of the wine tasted in the booths (8.14–8.61 vs. 7.45, respectively; Table 2). Conversely, the wine tasted in the real dark environment had a significantly lower astringency compared to that of the booths and the dark-VR environment (7.57 vs. 8.34–8.39, respectively; Table 2).

Virtual reality technology can provide consumers with simulated scenarios that are close to real environments [30]. The present research showed that changing the environment had a significant effect on perception, but that effect was marginal for acceptability. The overall liking of the wine sample in the real environments was similar compared to that of the VR environments. The same effect occurred for the perception of sweetness and acidity; however, there were significant differences between the real and VR in the dark environments for the perception of floral and astringency. Environments may affect consumers' expectations and experiences of products because their decisions can be unconsciously changed by several extrinsic factors [26]. Ryu and Jang [31] tested different contextual factors such as lighting, facility aesthetics, ambiance music, dining equipment, and employees' interactions on consumers dining experiences. They found that consumers showed positive emotions to simple environmental changes such as the type of music played and the layout of the dining environment. Moreover, the lighting conditions are very important for the sensory evaluation of foods. Bschaden et al. [32] found that the lighting conditions of the testing environment can affect the saltiness perception of tomato soups. Moreover, consumers tend to choose less healthy food options when the ambient lighting is dim [33]. In the present study, the dark environment might have been the most adequate contextual surrounding for consumers to taste the wine sample, as the perception of fruity and floral aroma had positive effects on consumers. In a similar study, Hersleth et al. [34] found that the wine tasting experience in a reception type of room was significantly improved compared to the tasting of the wines in traditional booths. With the development of more efficient virtual reality technologies, more sensory stimuli can be tested with different contextual situations. The virtual reality technology might potentially replace the use of physical environments in the future, becoming an important tool for sensory evaluation [35].

4.2. Emotions and Multivariate Analysis of the Wine Sample in Different Environments

In the present study, changes in the environment affected the elicited consumers' emotions toward the wine sample. Neutral to positive emotions such as "interested", "secure", "polite", and "friendly" were elicited for the wine tasted in the traditional booths and the real bright restaurant. The wine tasted in the real dark-restaurant and bright-VR environments was associated with "enthusiastic", "glad", and "aggressive". On the other hand, emotions such as "nostalgic" and "daring" were associated with the wine tasted in the dark-VR environment (Figure 3). Berridge and Kringelbach [36] stated that the environmental factors could change the human cognitive ability to elaborate the psychological representation of pleasant events, which might increase the perception of richness and taste by shaping the emotions that are felt toward the stimuli. This means that personal emotions are closely related to the environment in which they occur, and different consumption situations may have a significant effect on consumers' emotional responses [16].

The interaction between product and environment can affect the elicited emotions of consumers during the tasting. Piqueras-Fiszman and Jaeger [37] showed that the contextual scenario is, in fact, a trigger of emotional changes in the consumers. The scenarios that are considered more appropriate to consumers for food consumption had more positive elicited emotions compared to inappropriate contextual environments, which can produce more negative emotional terms associated with the product [37]. In the present study, the use of real restaurant environments triggered more positive elicited emotions compared to that of the traditional booths, which can be an indication of the appropriateness of the restaurant environments when tasting wine samples. In a similar study, Park and Farr [38] showed that consumers' emotions were affected by the lighting conditions of the testing environment. Changes in mood and emotions can also affect consumers' taste perception. In the present study, consumers perceived the wine sample to be higher in floral aroma and lower in astringency in the real dark-restaurant environment compared to that in the traditional booths. The real dark-restaurant environment was also responsible for generating positive emotions such as "enthusiastic", "glad", and "warm" when tasting the wine sample (Figure 3). Noel and Dando [39] showed that positive emotions were associated with increased sweetness and decreased acidity in ice-cream products, concluding that modulating taste perception can play an important role in emotional eating. In the present study, the overall liking of the wine sample was significantly and positively correlated with the perceived floral aroma and sweetness ($r = 0.3$ to 0.6), and negatively correlated with astringency ($r = -0.2$ to -0.4) for all testing environments (Table S1, Table S2, Table S3, Table S4 and Table S5).

In the present study, the purchase intent of the wine sample was marginally affected by the change in the environment. Consumers' decisions to buy a product are affected by several extrinsic factors such as packaging, logo, and color [40,41]. These external factors, combined with the contextual effects of the environment, can greatly modify the purchasing behaviors of consumers. Although traditional laboratory environments are designed to collect data, minimizing the influences of external contextual effects, these environments may lack ecological validity during the tasting. The use of real environments is an option to measure those external factors, but the experimental conditions might produce several variables that are difficult to control [42]. Virtual reality offers a novel solution to measure the effects of environmental factors and having controlled laboratory conditions.

Boesveldt et al. [43] stated that the development of the sensory perception is closely linked with the environment. Sensory perception plays an important role in the acceptance or rejection of food and drinks. Moreover, hedonic responses and preferences can be affected by the familiarity toward the food products that consumers have [28]. However, purchasing decisions of foods can be affected by the surrounding elements of different consumption environments, which means that each product can have a specific consumption environment that is suitable for it [37]. King et al. [44] found that the dining environment greatly affected the consumers' acceptance and choice of products. García-Segovia et al. [45] stated that the table decoration and dining place also affected consumers' acceptance and perception of foods. In the present study, the real environment had a greater effect on participants' perceptions of floral aroma and astringency compared to that of the VR environments. However, the emotions elicited by the wine sample were greatly affected by the VR environments; in particular, this effect was more evident for the dark environment. Future studies are needed to evaluate the effects of other environmental factors such as music and consumers' interactions with the use of novel virtual reality technologies.

5. Limitations

No re-tasting of test samples was possible under the VR testing process, which could be a limitation of this method and the results. Another limitation of this study was that participants were unable to assess the appearance of the sample when wearing the virtual reality headsets. Future technologies such as augmented reality (AR) can overcome this issue and provide a more immersive approach than VR; however, future studies are needed.

6. Conclusions

The context and environment affected the perceptual responses of consumers when tasting a wine product. Although liking was marginally affected, the dark-VR environment elicited different emotional responses compared to those of the traditional booths. The virtual reality technology provides a relatively stable and inexpensive method for sensory evaluation by providing a more realistic consuming environment compared to that of traditional sensory booths.

Supplementary Materials: The following are available online at http://www.mdpi.com/2304-8158/9/2/191/s1, Table S1: Correlation analysis (r)* of the variables measured for the wine sample in the traditional booth environment, Table S2: Correlation analysis (r)* of the variables measured for the wine sample in the bright restaurant environment, Table S3: Correlation analysis (r)* of the variables measured for the wine sample in the dark restaurant environment, Table S4: Correlation analysis (r)* of the variables measured for the wine sample in the bright VR environment, Table S5: Correlation analysis (r)* of the variables measured for the wine sample in the dark VR environment.

Author Contributions: Conceptualization, D.D.T. and Y.H.; data curation, D.D.T. and Y.H.; formal analysis, D.D.T. and Y.H.; funding acquisition, D.D.T., S.F., and F.R.D.; investigation, D.D.T., Y.H., C.S., S.F., C.G.V., and F.R.D.; methodology, D.D.T and Y.H.; project administration, D.D.T., S.F., and F.R.D.; resources, D.D.T., S.F., C.G.V., and F.R.D.; software, D.D.T., S.F., and C.G.V.; supervision, D.D.T, S.F., and F.R.D.; validation, D.D.T., S.F., C.G.V., and F.R.D.; visualization, D.D.T. and Y.H.; writing—original draft, D.D.T., Y.H., C.S., S.F., C.G.V., and F.R.D.; writing—review and editing, D.D.T., Y.H., C.S., S.F., C.G.V., and F.R.D. All authors have read and agreed to the published version of the manuscript.

Funding: This research was partially funded by the 2017 Early Career Researcher Grant Scheme from the University of Melbourne, Australia (603403), and the Australian Government through the Australian Research Council (IH120100053).

Conflicts of Interest: The authors declare no conflict of interest. The funders had no role in the design of the study; in the collection, analyses, or interpretation of data; in the writing of the manuscript; or in the decision to publish the results. The use of trade names does not imply endorsement, nor lack of endorsement of those not mentioned.

References

1. Spence, C. Multisensory experiential wine marketing. *Food Qual. Prefer.* **2019**, *71*, 106–116. [CrossRef]
2. Lesschaeve, I. Sensory evaluation of wine and commercial realities: Review of current practices and perspectives. *Am. J. Enol. Vitic.* **2007**, *58*, 252–258.
3. Schmidtke, L.M.; Rudnitskaya, A.; Saliba, A.J.; Blackman, J.W.; Scollary, G.R.; Clark, A.C.; Rutledge, D.N.; Delgadillo, I.; Legin, A. Sensory, chemical, and electronic tongue assessment of micro-oxygenated wines and oak chip maceration: assessing the commonality of analytical techniques. *J. Agric. Food Chem.* **2010**, *58*, 5026–5033. [CrossRef] [PubMed]
4. Vannier, A.; Brun, O.X.; Feinberg, M.H. Application of sensory analysis to champagne wine characterisation and discrimination. *Food Qual. Prefer.* **1999**, *10*, 101–107. [CrossRef]
5. Reynolds, A.G. *Managing Wine Quality: Viticulture and Wine Quality*; Elsevier: Amsterdam, The Netherlands, 2010.
6. Lattey, K.A.; Bramley, B.R.; Francis, I.L.; Herderich, M.J.; Pretorius, S. Wine quality and consumer preferences: understanding consumer needs. *Wine Ind. J.* **2007**, *22*, 31–39.
7. Stelick, A.; Dando, R. Thinking outside the booth—the eating environment, context and ecological validity in sensory and consumer research. *Curr. Opin. Food Sci.* **2018**, *21*, 26–31. [CrossRef]
8. Bangcuyo, R.G.; Smith, K.J.; Zumach, J.L.; Pierce, A.M.; Guttman, G.A.; Simons, C.T. The use of immersive technologies to improve consumer testing: The role of ecological validity, context and engagement in evaluating coffee. *Food Qual. Prefer.* **2015**, *41*, 84–95. [CrossRef]
9. Liu, R.; Hannum, M.; Simons, C.T. Using immersive technologies to explore the effects of congruent and incongruent contextual cues on context recall, product evaluation time, and preference and liking during consumer hedonic testing. *Food Res. Int.* **2019**, *117*, 19–29. [CrossRef]
10. Motoki, K.; Saito, T.; Nouchi, R.; Kawashima, R.; Sugiura, M. A sweet voice: The influence of cross-modal correspondences between taste and vocal pitch on advertising effectiveness. *Multisens. Res.* **2019**, *32*, 401–427. [CrossRef]
11. Spence, C.; Velasco, C.; Knoeferle, K. A large sample study on the influence of the multisensory environment on the wine drinking experience. *Flavour* **2014**, *3*, 8. [CrossRef]
12. Jackson, R.S. *Wine Tasting: A Professional Handbook*; Academic Press: Cambridge, MA, USA, 2016.
13. Motoki, K.; Saito, T.; Park, J.; Velasco, C.; Spence, C.; Sugiura, M. Tasting names: Systematic investigations of taste-speech sounds associations. *Food Qual. Prefer.* **2020**, *80*, 103801. [CrossRef]
14. Singh, A.; Beekman, T.L.; Seo, H.-S. Olfactory Cues of Restaurant Wait Staff Modulate Patrons' Dining Experiences and Behavior. *Foods* **2019**, *8*, 619. [CrossRef] [PubMed]
15. Fiegel, A.; Childress, A.; Beekman, T.L.; Seo, H.-S. Variations in Food Acceptability with Respect to Pitch, Tempo, and Volume Levels of Background Music. *Multisens. Res.* **2019**, *32*, 319–346. [CrossRef] [PubMed]

16. Dorado, R.; Chaya, C.; Tarrega, A.; Hort, J. The impact of using a written scenario when measuring emotional response to beer. *Food Qual. Prefer.* **2016**, *50*, 38–47. [CrossRef]

17. Lawless, H.T.; Heymann, H. *Sensory Evaluation of Food: Principles and Practices*; Springer Science & Business Media: Berlin/Heidelberg, Germany, 2010.

18. Sester, C.; Deroy, O.; Sutan, A.; Galia, F.; Desmarchelier, J.-F.; Valentin, D.; Dacremont, C. "Having a drink in a bar": An immersive approach to explore the effects of context on drink choice. *Food Qual. Prefer.* **2013**, *28*, 23–31. [CrossRef]

19. Nijman, M.; James, S.; Dehrmann, F.; Smart, K.; Ford, R.; Hort, J. The effect of consumption context on consumer hedonics, emotional response and beer choice. *Food Qual. Prefer.* **2019**, *74*, 59–71. [CrossRef]

20. Meilgaard, M.C.; Carr, B.T.; Civille, G.V. *Sensory Evaluation Techniques*; CRC press: Boca Raton, FL, USA, 1999.

21. Lu, S.-Y.; Shpitalni, M.; Gadh, R. Virtual and augmented reality technologies for product realization. *CIRP Ann.* **1999**, *48*, 471–495. [CrossRef]

22. Jaeger, S.; Hort, J.; Porcherot, C.; Ares, G.; Pecore, S.; MacFie, H. Future directions in sensory and consumer science: Four perspectives and audience voting. *Food Qual. Prefer.* **2017**, *56*, 301–309. [CrossRef]

23. Ammann, J.; Hartmann, C.; Peterhans, V.; Ropelato, S.; Siegrist, M. The relationship between disgust sensitivity and behaviour: A virtual reality study on food disgust. *Food Qual. Prefer.* **2020**, *80*, 103833. [CrossRef]

24. Goedegebure, R.P.; van Herpen, E.; van Trijp, H.C. Using product popularity to stimulate choice for light products in supermarkets: An examination in virtual reality. *Food Qual. Prefer.* **2020**, *79*, 103786. [CrossRef]

25. Pennanen, K.; Närväinen, J.; Vanhatalo, S.; Raisamo, R.; Sozer, N. Effect of virtual eating environment on consumers' evaluations of healthy and unhealthy snacks. *Food Qual. Prefer.* **2020**, 103871. [CrossRef]

26. Hannum, M.; Forzley, S.; Popper, R.; Simons, C.T. Does environment matter? Assessments of wine in traditional booths compared to an immersive and actual wine bar. *Food Qual. Prefer.* **2019**, *76*, 100–108. [CrossRef]

27. Ng, M.; Chaya, C.; Hort, J. Beyond liking: Comparing the measurement of emotional response using EsSense Profile and consumer defined check-all-that-apply methodologies. *Food Qual. Prefer.* **2013**, *28*, 193–205. [CrossRef]

28. Torrico, D.D.; Fuentes, S.; Viejo, C.G.; Ashman, H.; Dunshea, F.R. Cross-cultural effects of food product familiarity on sensory acceptability and non-invasive physiological responses of consumers. *Food Res. Int.* **2019**, *115*, 439–450. [CrossRef] [PubMed]

29. Bradley, M.M.; Lang, P.J. Measuring emotion: the self-assessment manikin and the semantic differential. *J. Behav. Ther. Exp. Psychiatry* **1994**, *25*, 49–59. [CrossRef]

30. Ung, C.-Y.; Menozzi, M.; Hartmann, C.; Siegrist, M. Innovations in consumer research: The virtual food buffet. *Food Qual. Prefer.* **2018**, *63*, 12–17. [CrossRef]

31. Ryu, K.; Jang, S.S. The effect of environmental perceptions on behavioral intentions through emotions: The case of upscale restaurants. *J. Hosp. Tour. Res.* **2007**, *31*, 56–72. [CrossRef]

32. Bschaden, A.; Dörsam, A.; Cvetko, K.; Kalamala, T.; Stroebele-Benschop, N. The impact of lighting and table linen as ambient factors on meal intake and taste perception. *Food Qual. Prefer.* **2020**, *79*, 103797. [CrossRef]

33. Biswas, D.; Szocs, C.; Chacko, R.; Wansink, B. Shining light on atmospherics: How ambient light influences food choices. *J. Mark. Res.* **2017**, *54*, 111–123. [CrossRef]

34. Hersleth, M.; Mevik, B.-H.; Næs, T.; Guinard, J.-X. Effect of contextual factors on liking for wine—Use of robust design methodology. *Food Qual. Prefer.* **2003**, *14*, 615–622. [CrossRef]

35. Jaeger, S.R.; Porcherot, C. Consumption context in consumer research: Methodological perspectives. *Curr. Opin. Food Sci.* **2017**, *15*, 30–37. [CrossRef]

36. Berridge, K.C.; Kringelbach, M.L. Affective neuroscience of pleasure: reward in humans and animals. *Psychopharmacology* **2008**, *199*, 457–480. [CrossRef] [PubMed]

37. Piqueras-Fiszman, B.; Jaeger, S.R. The impact of evoked consumption contexts and appropriateness on emotion responses. *Food Qual. Prefer.* **2014**, *32*, 277–288. [CrossRef]

38. Park, N.K.; Farr, C.A. The effects of lighting on consumers' emotions and behavioral intentions in a retail environment: A cross-cultural comparison. *J. Inter. Des.* **2007**, *33*, 17–32. [CrossRef]

39. Noel, C.; Dando, R. The effect of emotional state on taste perception. *Appetite* **2015**, *95*, 89–95. [CrossRef] [PubMed]

40. Torrico, D.D.; Fuentes, S.; Viejo, C.G.; Ashman, H.; Gurr, P.A.; Dunshea, F.R. Analysis of thermochromic label elements and colour transitions using sensory acceptability and eye tracking techniques. *LWT* **2018**, *89*, 475–481. [CrossRef]

41. Luffarelli, J.; Stamatogiannakis, A.; Yang, H. The visual asymmetry effect: An interplay of logo design and brand personality on brand equity. *J. Mark. Res.* **2019**, *56*, 89–103. [CrossRef]

42. Meiselman, H.L. The role of context in food choice, food acceptance and food consumption. *Front. Nutr. Sci.* **2006**, *3*, 179.

43. Boesveldt, S.; Bobowski, N.; McCrickerd, K.; Maître, I.; Sulmont-Rossé, C.; Forde, C.G. The changing role of the senses in food choice and food intake across the lifespan. *Food Qual. Prefer.* **2018**, *68*, 80–89. [CrossRef]

44. King, S.C.; Meiselman, H.L.; Hottenstein, A.W.; Work, T.M.; Cronk, V. The effects of contextual variables on food acceptability: A confirmatory study. *Food Qual. Prefer.* **2007**, *18*, 58–65. [CrossRef]

45. García-Segovia, P.; Harrington, R.J.; Seo, H.-S. Influences of table setting and eating location on food acceptance and intake. *Food Qual. Prefer.* **2015**, *39*, 1–7. [CrossRef]

MDPI
St. Alban-Anlage 66
4052 Basel
Switzerland
Tel. +41 61 683 77 34
Fax +41 61 302 89 18
www.mdpi.com

Foods Editorial Office
E-mail: foods@mdpi.com
www.mdpi.com/journal/foods

* 9 7 8 3 0 3 6 5 2 5 3 7 2 *